Silicon Second Nature

SILICON

SECOND

NATURE

Culturing Artificial Life

in a Digital World

STEFAN HELMREICH

UNIVERSITY OF CALIFORNIA PRESS

Berkeley · *Los Angeles* · *London*

University of California Press
Berkeley and Los Angeles, California

University of California Press, Ltd.
London, England

© 1998 by
The Regents of the University of California

Library of Congress Cataloging-in-Publication Data

Helmreich, Stefan, 1966–
   Silicon second nature : culturing artificial life
in a digital world / Stefan Helmreich.
   p.   cm.
   Includes bibliographical references and index.
   ISBN 0-520-20799-8 (alk. paper)
   1. Biological systems—Computer simulation—Research.
   2. Biological systems—Computer simulation—Philosophy.
   3. Santa Fe Institute (Santa Fe, N.M.) I. Title.
QH324.2.H45 1998   98-19393
570'.1'13—dc21   CIP

Printed in the United States of America
9 8 7 6 5 4 3 2 1

Contents

Acknowledgments

FOLLOWING THE SCIENCE of Artificial Life requires a kind of hyper-textual travel through dense thickets of connection between machines, institutions, and people. In writing this book I have been guided by a constellation of Artificial Life researchers, anthropological colleagues, and friends and loved ones. My first webs of gratitude must gather together the scientists who aided me in mapping the territories that link Artificial Life to social life. I am indebted to Steen Rasmussen for first hearing and encouraging my interest in studying among the people at the Santa Fe Institute. Walter Fontana opened my eyes to the meshworks of metaphor that make worlds out of computers, Brian Goodwin traced out the patterns that make organisms whole, Terry Jones instructed me in the ricocheting ways of the Echo system, Ken Karakotsios took me behind the screens of SimLife, Chris Langton engaged anthropology as an ally in the search for novel ways of theorizing "life," Melanie Mitchell tested the fitness of my writings on evolutionary recipes in computer science, Una-May O'Reilly taught me how to see nature in the nodes of the Internet, John Stewart mentored me in the historical materialism of autopoiesis, and Larry Yaeger made information theory hum with a supernatural scientific animism. Other Artificial Life and Santa Fe Institute researchers who steered me through the interleaved landscapes of science and society were David Ackley, Mark Bedau, Philippe Binder, John Casti, Claus Emmeche, Arantza Etxeberria, Doyne Farmer, Julio Fernández Ostolaza, Stephanie Forrest, Valerie Gremillion, Inman Harvey, George Kampis, Brian Keeley, John Koza, Dominique Lestel, Geoffrey Miller, Nelson Minar, Eric Minch, Mats Nordahl, Richard Palmer,

Mukesh Patel, Tom Ray, Craig Reynolds, Bruce Sawhill, Becca Shapley, Karl Sims, Joshua Smith, Chuck Taylor, Kurt Thearling, Peter Todd, Francisco Varela, and David Wolpert, along with many others whose words also traverse this text. I thank Santa Fe Institute vice president Mike Simmons and program director Ginger Richardson, who arranged for me to be accepted in the corridors at Santa Fe.

The members of my dissertation committee at Stanford University helped me immensely in articulating the first bones of this book. I thank Jane Collier, for skeleton-shaking criticism delivered with a sense of liberal humor and history; Carol Delaney, for lifesaving transfusions of feminist theories of creation and procreation; William Durham, for tonic attention to biological detail and a sympathetic skepticism; Joan Fujimura, for careful counsels on how to operate amid the politics of laboratory life; Akhil Gupta, for incisive dissections of the ever-adapting body of advanced capitalism; and Timothy Lenoir, for unremitting enthusiasm for my idiosyncratic ways of figuring science as culture. Courses at Stanford with George Collier, Barry Katz, Andrea Klimt, and Terry Winograd also animated my thinking on Artificial Life at early stages. I thank the fellows at the Stanford Humanities Center, where I was in residence as I wrote the dissertation; the members of the Department of Science and Technology Studies at Cornell University, where I began drafting the book; and the fellows at the Center for the Critical Analysis of Contemporary Culture at Rutgers University, where I finished the work.

I am grateful to people in the distributed community of cultural studies of science. Particular thanks must go to Julian Bleecker, for sharing suspicions of simulation games; Marianne de Laet, for talking of travel and transubstantiation in the spheres of science and sentiment; Rich Doyle, for folding my anthropological algorithms into Möbius conformations; Joe Dumit, for mindful commentary on early slices of the book; Paul Edwards, for wise words on microworlds; Sarah Franklin, for asking me to culture my arguments in unexpected theoretical media; Hugh Gusterson, for advice on how to handle radioactive scientists; Donna Haraway, for teaching me about the technoscientific time machine called the twentieth century; N. Katherine Hayles, for reversing figure and ground in Artificial Life; Deborah Heath, for feminist reinforcement in a variety of cyberspaces; and Mimi Ito, for key conversations about the realities that people create in computers. My science studies net was also knit together by Karen Barad, Monica Casper, Giovanna Di Chiro, Gary Downey, Ron Eglash, Arturo Escobar, Michael Fischer, Anne Foerst,

Peter Galison, Saul Halfon, Cori Hayden, Linda Hogle, Evelyn Fox Keller, Chris Kelty, Michael Lynch, Emily Martin, David Noble, Bryan Pfaffenberger, Trevor Pinch, Rayna Rapp, Jennifer Reardon, Hans-Jörg Rheinberger, Osamu Sakura, Sahotra Sarkar, Phoebe Sengers, Neil Smith, Karen-Sue Taussig, Peter Taylor, Sharon Traweek, Sherry Turkle, and Norton Wise. The late Diana Forsythe was a compassionate comrade in the anthropology of computing; her commitments to connecting the personal, political, and digital have informed my own work deeply.

I am indebted to friends and colleagues for rich compounds of intellectual and moral support. I thank Asale Ajani, whose writings on lineage, hybridity, race, twinning, and aliens forced me to rethink identity; Samer Alatout, who spirited me off to New Orleans for Mardi Gras speculations on Foucault; Susan DeLay, who saved me when I was snowed under by lazy blue satans in Ithaca; David Derrick, who routed me through the tangled circuits of airports and electronic musical instruments; Jack Ferguson, who tutored me in transmitting my own life history; Helen Gremillion, who shared a persistent wondering about how "nature" came to be anchored in flesh; Rosalva Aída Hernández, whose friendship took me across contexts in the United States, Mexico, and elsewhere; Chris Henry, who built a smiley-faced thamtor with me during a fractured week in Santa Fe; Kiersten Johnson, whose thoughts on the rhetoric and mystical tones of Artificial Life gave me emotional and intellectual sustenance during the research and writing of the dissertation; Debra Lotstein, who taught me about the responsibilities of knowing in religion and biology in Jerusalem and Palo Alto; Lisa Lynch, who began to reconfigure my words and worlds in Manhattan; Bill Maurer, who started me thinking about what "nature" meant anyway; Saba Mahmood, who alerted me to the religious tones of the secular over uncommon meals and email from Cairo to California; Cris Moore, who filled my Santa Fe days with Green politics, word and number puzzles, computational theosophies, and frenzied backgammon games; Diane Nelson, who wrote and acted with dead serious science fictional levity; Heather Paxson, who gave me wisdom and friendship during my most turbulent times and who first coaxed me to think about new reproductive technologies alongside Artificial Life; Dmitry Portnoy, who sent me biblical commentaries in the form of screenplays; Marcelle Poulos, who schooled me in radical pedagogy and songwriting in Los Angeles and San Francisco; Lucía Rayas, whose devotion to the poetry of translation helped me transmute programs into prose; Sarah Richards-Gross, who astounded me with stories about how to conduct experiments underwater; Lars Risan, whose experience doing the anthropology of Artificial Life in the United Kingdom was an invitation to intellectual fencing and friendship; Nikolai Ssorin-Chaikov, who made me think about social reality backward; Liliana Suárez Navaz, who shared experiences and theories of nomadism with me in the United States and in Andalucía; Rebecca Underwood, who taught me much of what I know about computers and about the technical and emotional acrobatics involved in maintaining telecommunicative friendships; Miranda von Dornum, who showed me the wonders of the polymerase chain reaction and of cottontop tamarins at Harvard; and Al-Yasha Williams, who cornered me on color theory. Also threaded through this book are inspirations from Suraj Achar, Wendy Anderson, Ann Bell, Genevieve Bell, Federico Besserer, Simon Cheffins and people of Crash Worship, Jenny Cocq, Linda Feferman, Sheila Foster, Brenda Goodell, the Gutierrez family, Samira Haj, Jon Haus, Dixie Hellfire Katzenjammer, Ayse Koktvedgaard/Miriam Zeitzen, Jennifer Lichtman, Cathy Lindgren, Alejandro Lugo, Melissa McDonald, Colleen O'Neill, Amanda Pascall, Tal Raz, Kamran Sahami, Suzana Sawyer, Mukund Subramanian, David Wine, and Yatsu.

Thanks to my close biogenetic and other rhizomatic kin, especially my parents, Mary and Gisbert Helmreich, who taught me to appreciate the aesthetic angles of technology, the mathematical magic of chess, and the geometries of southern California beaches, and who gave me love that has been more important than they know. My paternal grandmother, Dorothea Helmreich, in Germany, looked after me in a language I never mastered, and my maternal grandmother, Erdna Rogers, woke me up to the wondrous waters and woods of New England, even when I wanted to stay indoors and read. My mother's siblings, Ann, Peter, Mark, and Lucy Rogers, have been present at crucial moments, as have my matrilateral parallel cousins, Phoebe and Amos Hausmann-Rogers, my matrilateral cross cousins, Amy and Robbie Rogers, and my uncle Mark's partner, Sally Campbell. The memory of my grandfather Howard Rogers visits me often, for it was he who first educated me about optics and the peculiar nature of stereoscopy.

The staff at the Santa Fe Institute—particularly Marita Prandoni, Fritz Adkins, Patrisia Brunello, Ronda Butler-Villa, Deborah Smith, and Scott Yelich—were ever helpful. Institute librarian Margaret Alexander was a spring of thoughtful humor, moral support, and bibliographic guidance. In Stanford Anthropology, the administrative aid of Beth Bashore,

Jeanne Giaccia, and Shannon Brown was essential. At the Stanford Humanities Center, I must acknowledge Gwen Lorraine, Susan Sebbard, and Sue Dambrau; at Cornell, Debbie van Gelder, Marta Weiner, and Cindy Dougherty; and at Rutgers, Vanessa Ignacio and Link Larsen. My editors at the University of California Press, particularly Stanley Holwitz, Scott Norton, Barbara Jellow, and Sheila Berg, were instrumental in getting this book out of its petri dish and into the ecology of the publishing world. Mary Murrell at Princeton University Press made healthy suggestions at an early stage.

Financial support for my fieldwork was provided by a grant from the National Science Foundation (SBR-9312292) and through a Predoctoral Research Assistantship granted through the Stanford University Anthropology Department. Support during the process of thesis writing came from a Mellon Dissertation Writeup Grant administered through the Department of Anthropology, from a Melvin and Joan Lane Graduate Fellowship in the History of Science, and from a Dissertation Resident Fellowship provided through the Stanford Humanities Center. A postdoctoral associateship at Cornell University gave me money to go to Japan and a Rockefeller Foundation Humanities Fellowship at Rutgers University helped me sew together the final body of this text, sealing this bond to the lives of friends and colleagues near and far, here and now, past and future.

Arguments I originally developed in the following articles are reprinted with permission: "Replicating Reproduction in Artificial Life: or, the Essence of Life in the Age of Virtual Electronic Reproduction," in *Reproducing Reproduction: Kinship, Power, and Technological Innovation,* ed. Sarah Franklin and Helena Ragoné, 207–234 (Philadelphia: University of Pennsylvania Press, 1998); "Recombination, Rationality, Reductionism, and Romantic Reactions: Culture, Computers, and the Genetic Algorithm," *Social Studies of Science,* 28, no. 1: 39–71 (Copyright Sage Publications Ltd., 1998); "The Spiritual in Artificial Life: Recombining Science and Religion in a Computational Culture Medium," *Science as Culture* 6, part 3 (1997): 363–395.

Epigraphs to the following chapters are reproduced with permission:

Introduction, from *The Cyberiad: Fables for the Cybernetic Age,* by Stanislaw Lem. English translation copyright © 1974 by Seabury Press. Reprinted by permission of Continuum Publishing Company.

Chapter 1, from *Slaves of the Machine: The Quickening of Computer Technology*, by Gregory J. E. Rawlins. Copyright © 1997 by MIT Press. Reprinted by permission of MIT Press/Bradford Books.

Chapter 2, from *Gödel, Escher, Bach: An Eternal Golden Braid*, by Douglas Hofstadter. Copyright © 1979 by Basic Books. Reprinted by permission of HarperCollins Publishers, Inc.

Chapter 4, from *Zen and the Art of Motorcycle Maintenance: An Inquiry into Values*, by Robert M. Pirsig. Copyright © 1974 by Robert M. Pirsig. Reprinted by permission of William Morrow and Company, Inc.

Chapter 5, from *Written on the Body*, by Jeanette Winterson. Copyright © 1992 by Great Moments, Ltd. Reprinted by permission of Random House, Inc.

Santa Fe, New York City, and San Francisco, 1997

He built a machine and fashioned a digital model of the Void, an Electrostatic Spirit to move upon the face of the electrolytic waters, and he introduced the parameter of light, a protogalactic cloud or two, and by degrees worked his way up to the first ice age.

STANISLAW LEM, *The Cyberiad*

Introduction

IN THE BEGINNING, Tom Ray created Tierra, an elementary computer model of evolution. Laboring late into the night in early January 1990, Ray released a single self-replicating program into a primordial information soup he had programmed, producing what he would come to call a computational "ecosystem" in which "populations" of "digital organisms" could "evolve." Ray happily extended words associated with life to this "artificial world" because he defined evolution as the story of the differential survival and replication of information structures. For Ray and many others in the nascent scientific field of Artificial Life, computer programs that self-replicate—like computer viruses—can be considered new forms of life, forms that can be quickened into existence by scientists who view the computer as an alternative universe ready to be populated with reproducing, mutating, competing, and ultimately unpredictable programs.

On July 7, 1994, some four years after Tierra's nativity, Ray spoke to a large audience of computer scientists, biologists, and engineers at a Massachusetts Institute of Technology (MIT) conference on Artificial Life. He suggested that the digital organisms in Tierra needed more space to evolve. He proposed that Tierra be expanded, that people using computer networks around the world volunteer to accept a franchise of the system, that they give a portion of their Internet accounts over to running Tierra as a "low-priority background process," that Tierra—Spanish for "Earth"—become coextensive with our planet. Ray wanted the habitat of Tierran organisms to be jacked up from a single computer memory to the memories of many machines the world over, from a space the informatic equivalent of a drop of water to one the equivalent of a small pond. Ray's talk, "A Proposal to Create a Network-Wide Biodiversity Reserve for Digital Organisms," was impassioned, almost evangelical. As he paced the stage, Ray explained that the Tierran ecology could only really blossom if it could be expanded into global cyberspace. Only in this way might there be anything analogous to the Cambrian explosion of diversity in the organic world, only in this way might self-replicating computer programs evolve into software creatures that might be harvested, domesticated, and bred for potentially useful applications. Ray's vision was spectacular: he hoped that Tierran organisms could "roam freely" in a cyberspace reserve, traveling around the globe in search of spare central processing unit (CPU) cycles, likely following nightfall as cycles were freed up by humans logging off for the day. Ray stopped pacing and said, "I think of these things as alive, and I'm just trying to figure out a place where they can live." He mentioned a parallel project to consolidate a nature reserve in Costa Rica, where he began his career as a tropical ecologist: "I'm doing the same thing in the tropical rain forest. I sort of see these two projects as conceptually the same." Ray's conviction that he had authored a new instance of life motivated his pleas to the audience to participate in his project, to, as he put it repeatedly, "give life a chance."

Ray was not alone in his belief that he had created life in a computer. The MIT conference was populated by many scientists who believed that programs could count as life-forms, or, at the very least, as models of life-forms. Just a few hours after Ray's talk, researchers gathered in a capacious computer simulation demonstration hall, where they moved in flocks from one computer screen to the next, waiting for the expectant scientist stationed at each machine to say something about the artificial world he or she had created. Behind the glowing glass screens of Macintoshes, Sun Sparcstations, and Silicon Graphics Iris workstations hovered images of colorful artificial fish, pictures of roving two-eyed Ping-Pong ball–shaped creatures, and odd triangular and trapezoidal figures that chased each other around on a planar surface. Through some screens researchers could look down over imaginary landscapes where shifting patterns of dots represented populations of elementary organisms competing over territory and resources. A few simulations presented the viewer only with ever-updating graphs of population statistics for self-replicating programs.

Most simulations were designed for didactic and experimental purposes, as illustrations or abstract models of the dynamics of evolution and lifelike behavior in populations. Such artifacts often consisted of large programs containing packs of smaller self-replicating programs meant to represent populations of real creatures, such as ants, rabbits, or mosquitoes. Artificial Life researchers dedicated to these sorts of models were convinced that through distilling the logic of evolution in a computer, they might hatch ideas for a theoretical biology that could account for both real and possible life. Some other scientists, like Ray, went beyond such modest claims, maintaining that computer processes exhibiting suitably lifelike behavior could be considered new instances of life itself. They said that their self-reproducing algorithms were *real* artificial life-forms; no mere representations or counterfeits of life, these algorithms were artifactual creatures ultimately realized as material entities in the voltage patterns deep within computers. Researchers responsible for these sorts of programs hoped that through creating swarms of self-replicating entities in a virtual universe, they might add to the dominion of life a new kingdom of organisms existing in the universe of cyberspace.

The claims of some Artificial Life researchers to have synthesized life may sound strikingly novel, but they also mutate a well-rooted historical tradition of attempting to manufacture living things, a tradition that entwines activities that have been variously mystical, literary, religious, technological, and scientific. The Pygmalion myth tells of a sculptor who made an ivory statue, Galatea, with which he fell hopelessly in love, and which, with the help of Venus, he brought to life with a kiss. Talmudic lore tells of a Rabbi Löw of Prague who, in 1580, fashioned a creature of clay called the Golem, which he brought to life by breathing into its mouth the ineffable name of God, an act that appropriated the divine creative power of the Word. Mary Shelley's *Frankenstein* (1818) famously sets forth the tale of a jigsaw creature jolted to life with electricity. Shelley cast Doctor Frankenstein as a modern Prometheus, that figure in Greek mythology who stole fire from the gods and created humanity from wet earth. In *Faust*, Goethe wrote of a young student who made a little man in a vial, a creature whose first words enunciate some of the themes of unnatural fatherhood and supernatural fear that have attended the quest for artificial life. The homunculus speaks to his creator from behind a glass barrier prefiguring the computer screens that separate Artificial Life programmers from their creations:

HOMUNCULOUS (*speaking to Wagner from the phial*).
Well, Father, what's to do? No joke I see.
Come, take me to your heart, and tenderly!
But not too tight, for fear the glass should break.
That is the way that things are apt to take:
The cosmos scarce will compass Nature's kind,
But man's creations need to be confined.

                                              Goethe 1832:101

In these fanciful tales, life is synthesized through a sort of masculine birthing, a reproduction with no need for women's bodies, a reproduction that brings inert matter to life with a kiss, a breath, a word, a spark.

Synthetic life has been a grail for scientific theory and practice as well. Using the technology of clockwork mechanism in place of more archaic hydraulic and pneumatic techniques, people built a variety of automata during the Renaissance and the Enlightenment, the most famous of which played music or mimicked animal behavior. In 1748, Julien Offray de La Mettrie, in his *L'Homme machine*, argued against René Descartes's separation of mind and body and declared that all aspects of human vitality could be mechanized. In 1872, Samuel Butler reasoned in his book *Erewhon* that Darwinism conceived organisms as machines, opening up the possibility of machine life and evolution. By 1948, Norbert Wiener, a founding figure in cybernetics, was able to theorize animals and machines as kindred kinds of information-processing devices. Working on allied notions at around the same time, the mathematician John von Neumann proposed that machines hosting stored programs might be capable of reproducing themselves if such programs contained self-descriptions (Watson and Crick's later explanation of DNA's structure and function in 1953 in fact used the rhetoric of programming to suggest that DNA was a coded self-description folded up in organisms). In 1956, computer scientists gathered at Dartmouth University to establish the field of Artificial Intelligence, an endeavor aimed at making minds out of computers. A rich history has prepared the way for Artificial Life to make sense.

Tom Ray's research builds on this history. And at the MIT Artificial Life conference, Ray was a figure of some importance, recognized as one of the first to have successfully put together a simulation of evolution, to have moved fully from a view of the computer as substitute mind to one envisioning it as a surrogate world. His affiliation with the Santa Fe Institute for the Sciences of Complexity, a research center in Santa Fe,

New Mexico, devoted to the computer simulation of nonlinear phenomena, also gave his words a certain weight. The Institute is widely known as a site for innovative work in complexity science and as an epicenter of Artificial Life research. Ray's talk was much anticipated and his own excitement about his project was indexed by the mantric, John Lennon-like "give life a chance" chorus that punctuated his talk and that tagged him as part of a generation of 1960s and 1970s young adults grown into 1990s scientists.

Ray's injunction echoed off the walls of the lecture hall as I readied myself for my turn on stage. I had come to this conference as an anthropologist fascinated with the practices of Artificial Life and had just flown in from New Mexico, where I had finished up a year of fieldwork at the Santa Fe Institute. As I moved toward the podium, my life in Santa Fe flashed before me in an Adobe Photoshop blur. I remembered interviews with scientists arguing passionately that, yes, computers are alternative universes in all senses that matter; that, yes, life really is just information processing; and that, yes, evolution has selected an elite corps of computer scientists to facilitate its phase transition from carbon to silicon. But by now the story I had become interested in telling about Artificial Life was not one that celebrated it as some transcendent next evolutionary step. Rather, it was an anthropological tale, one interested in how people have come to think of computer programs as lifeforms and one curious about the practical, institutional, cultural, political, and emotional dimensions of Artificial Life work. It was a tale aimed at understanding how Artificial Life might herald new conceptions and configurations of the natural, the artificial, and the organic in late-twentieth-century U.S. and European culture. It was a story about the changing meaning of "life."

And it was a story I started to tell at this conference, which marked the conclusion of my extended fieldwork among Artificial Life scientists and the beginning of the process of writing this book, which is an ethnographic portrait of the Artificial Life community, especially that segment located at the Santa Fe Institute. Because the Artificial Life community extends beyond the Institute into a network of universities around the United States and Europe, this book also reports on interviews I carried out over electronic mail and on fieldwork I did at international conferences. Before I descend into the anthropological account that my fieldwork produced, though, let me rewind to locate Artificial Life on the scientific map, to say more about the field's origins, mission, institutional contexts, and technological attachments. Let me also set my own theoretical and methodological frames in place.

ARTIFICIAL LIFE

Artificial Life is a field largely dedicated to the computer simulation—and, some would ambitiously add, synthesis in real and virtual space—of biological systems. It emerged in the late 1980s, out of interdisciplinary conversations among biologists, computer scientists, physicists, and other scientists. Artificial Life researchers envision their project as a reinvigorated theoretical biology and as an initially more modest but eventually more ambitious enterprise than Artificial Intelligence. Whereas Artificial Intelligence attempted to model the mind, Artificial Life workers hope to simulate the life processes that support the development and evolution of such things as minds. They plan to capture on computers (or, sometimes, in autonomous robots) the formal properties of organisms, populations, and ecosystems. A mission statement on Artificial Life generated by the Santa Fe Institute summarizes the approach:

> Artificial Life ("AL" or "ALife") studies "natural" life by attempting to recreate biological phenomena from first principles within computers and other "artificial" media. Alife complements the analytic approach of traditional biology with a synthetic approach in which, rather than studying biological phenomena by taking apart living organisms to see how they work, researchers attempt to put together systems that behave like living organisms. Artificial life amounts to the practice of "synthetic biology." (Santa Fe Institute 1994b:38)

The conceptual charter for this practice of synthesizing new life is captured by the Artificial Life scientist Christopher Langton's declaration that life "is a property of the *organization* of matter, rather than a property of matter itself" (Langton 1988:74). Some have found this claim so compelling that they maintain that alternative forms of life can exist in computers, and they hope the creation of such life-forms can expand biology's purview to include not just *life-as-we-know-it* but also *life-as-it-could-be*—life as it might exist in other materials or elsewhere in the universe (Langton 1989:1). Although Artificial Life remains peripheral to mainstream biology, researchers are attempting to build institutional and interdisciplinary alliances, and they have generated enthusiasm among a few prominent figures in evolutionary biology.

The field was officially named in 1987, when Langton, then a post-doctoral fellow at Los Alamos National Laboratories, in Los Alamos, New Mexico, hosted a conference to explore how computers might be used to model biological systems. Langton gathered a group of people that included computer scientists, biologists, physicists, and philosophers, and he took the opportunity to christen this new field "Artificial Life," a move that would have a great impact on how the field would be advertised, organized, and historically situated as well as on how people would craft cross-disciplinary and international ties. One person summarized the move to me thus: "Artificial Life is an excellent phrase. It provokes attention. Without the phrase there would be no field of research, of that I am convinced." One young scientist interested in using simulations to model problems in animal and human evolution said to me, "I think, as a term, 'artificial life' is a stroke of advertising genius. A research field is a product that scientists market to governmental funding agencies, to prospective students, and to the general public. The more evocative the name, the better exposure the field gets. The name might not mean much formally, but its poetic power is undeniable." In the late 1980s and early 1990s, Artificial Life became a magnetic topic in popular science journalism because of its spectacular promises of creating new life and because its name suggested it as a successor to Artificial Intelligence. One older scientist, not centrally interested in Artificial Life, complained to me about the grandness of the name: "What an ill-defined subject, Artificial Life. A silly name. We live in an age of soundbites." Soundbites, however, can often become nutrients for serious research projects. At the end of the 1980s, Artificial Life became one of the research foci of Los Alamos's nearby relative, the Santa Fe Institute for the Sciences of Complexity.

In its literature, the Santa Fe Institute describes itself as a "private, independent organization dedicated to multidisciplinary scientific research and graduate education in the natural, computational, and social sciences" (Santa Fe Institute 1991:2).[1] Since its founding in 1984 by a confederation of older scientists working mostly at Los Alamos, the Institute has become a gathering ground for an international community interested in "complexity" and nonlinear dynamics in physical, chemical, biological, computational, and economic systems. The Institute regularly sponsors interdisciplinary workshops and serves as a central node in many research networks, and it has been instrumental in organizing the

U.S. conferences on Artificial Life that have made the field a going concern. The Institute is unique as a scientific research center in that the only
scientific devices in evidence are the many computers on which people do
the work of simulating complex systems. It is the use of simulation as a
common tool that facilitates the interdisciplinary interactions the Institute is so interested in fostering. Like Artificial Life, the Santa Fe Institute has received a good deal of celebratory press and has been featured
in many popular science forums including *Discover, Omni, Science,* and
the *New York Times.*

The Santa Fe Institute is only one of several sites around the world
where work in Artificial Life is conducted, or, perhaps more accurately,
where computer modeling of biological systems is done. This activity has
been under way in many places for quite some time.[2] A good portion of
Artificial Life work has to do with developing robots that act au-
tonomously and adaptively and that employ control systems developed
using insights from evolution. In this book, I focus most sharply on the
use of computer simulations to model and create "artificial worlds": virtual, alternative, sites for evolution.[3] I am concerned with simulation approaches partly because they predominated at Santa Fe, but also because
they carry the view of life as information processing to its most vivid
conclusion. Simulation also ushers scientific practice and theory into new
epistemological territory, territory that reshapes how scientists think
about the fit between theory and experiment and between representation and reality.

Not all Artificial Life scientists are happy with how the recent history
of the field is told, with how this shapes the terrain of inquiry, or with
how the Santa Fe Institute is privileged in popular accounts (particularly
in Levy 1992b; Lewin 1992; Waldrop 1992; Kelly 1994; Turkle 1995).
Many Europe-based researchers argue that Artificial Life was not created ex nihilo in New Mexico but has descended from tangled international lineages of cybernetics, systems theory, Artificial Intelligence,
self-organization theory, origins of life research, and theoretical biology.
The Chilean-born and Paris-based biologist Francisco Varela, known for
his theoretical work in cognitive science and immunology, has been quite
vocal about this (see Varela 1995). He has argued that the materializa-
tion of Artificial Life in New Mexico has focused attention on overly
computational views of life and that the naming of Artificial Life on the
analogy to Artificial Intelligence has only made this more intense. U.S.
narrations of Artificial Life history are notorious in the international

community for their erasure or marginalization of European and Latin American precedents and scientists, of transnational collaborations, and of the social, intellectual, and economic contexts that produced the science in some places and not others. Indeed, my own decision to do fieldwork on the community at Santa Fe was powerfully guided by readings of popular science, and this book runs the risk of reinforcing the mainstream myth that "Artificial Life was born out of Zeus' head in Santa Fe, New Mexico, in the 1980s"—as one Europe-based scientist sardonically summarized it to me.[4]

CULTURING ARTIFICIAL LIFE

Artificial Life is more than a new way of thinking about biology. It is a symptom and a source of mutating visions of "nature" and "life" in an increasingly computerized world. As such, it has a social, cultural, and historical specificity. This book is concerned with chasing down that specificity. I document how the local knowledges and artifacts of Artificial Life are produced in the institutional and imaginative spaces of the Santa Fe Institute and in the clusters of computer simulation that it holds in its orbit. My argument is that Artificial Life scientists' computational models of "possible biologies" are powerfully inflected by their cultural conceptions and lived understandings of gender, kinship, sexuality, race, economy, and cosmology and by the social and political contexts in which these understandings take shape. Ideas and experiences of gender and kinship circulating in the heterosexual culture in which most researchers participate, for example, inform theories about "reproduction," "sex," "relatedness," and "sexual selection" in artificial worlds, and notions of competition and market economics in the capitalist West shape the construction of "artificial ecologies" in which populations of programs vie to "survive" and "reproduce."

"Silicon second nature" is my name for the substance and space that Artificial Life researchers seek to create in computers. The concept of second nature has a lineage traceable to Hegel, who "taught us to see a difference between 'first nature'—the given, pristine, edenic nature of physical and biotic processes, laws, and forms—and 'second nature.' Second nature comprises the rule-driven social world of society and the market, culture and the city, in which social change is driven by a parallel set of socially imposed laws"(Smith 1996:49). Artificial Life worlds are second natures in that they are rule-ordered human constructions, but also in that they are meant to mirror first nature. And they are second natures in still another way: they not only ape first nature but also offer to replace it, to succeed it as a resource for scientific knowledge. Second nature, of course, also refers to human habits and cultural practices that, through repetition, come to seem rooted in organic common sense. As Artificial Life researchers embrace the logics of synthetic vitality, they come to possess a new sort of subjectivity, a silicon second nature that may be increasingly common among humans inhabiting a world in which computers are haunted by "life."

This does not mean that the configurations precipitating from the solutions of Artificial Life are entirely new. Artificial life and artificial worlds are cultured in a social medium located in the powerful history of Western technoscience. More often than not, the first nature that sits as a model for silicon second nature is marked by images of creation, individuals, lineages, families, economies, and communities resonant with the values and practices of white middle-class people in secularized Judeo-Christian cultures in the United States and Europe. Of course, clean lines cannot be drawn from the demographic profile of the Artificial Life community to ideologies or imagery encoded in their computational work. It is messier than that. But not so complex that regularities cannot be discerned. I am interested in "hegemonic" rather than neatly "ideological" stories, in stories that find sustenance in pervasive commonsensical, almost unconscious, dominant ways of understanding, experiencing, and acting in the world (see Gramsci 1971; Williams 1977; Comaroff and Comaroff 1991). I am not always or solely concerned with the motives of Artificial Life researchers. In looking at imagery encoded in simulations, I follow the feminist theorist Carol Cohn, who has argued that "individual motivations cannot necessarily be read directly from imagery; the imagery itself does not originate in these particular individuals but in a broader cultural context" (1987:693).

This is not to say that people's motives are unimportant—at many moments I *am* concerned with researchers' stated inspirations—only that they are not the whole story. Because much of this book is devoted to analyses of simulations, I tend to foreground the role of cultural narratives in rendering Artificial Life artifacts legible to the scientific communities that consume them. I assume that science, while institutionally and discursively set apart from other human practices, crucially intersects with and is fundamentally constituted by ostensibly nonscientific activities and ideas. Following the anthropologist Marilyn Strathern, I understand "culture" to consist in part "in the way people draw analogies between different domains in their worlds" (1992b:47). The practice of Artificial Life is fully cultural in this sense; different domains— cultural and scientific, natural and artificial—are drawn together to create new ways of theorizing life. Twisting Chris Langton's formulation to my own purposes, I contend that constructions of *life-as-it-could-be* are built from culturally specific visions of *life-as-we-know-it*.

The logic of Artificial Life promises to transform the texture of our everyday experience of machines and programs. We already encounter computers as beyond our comprehension, as animated by logics that make them opaque, even uncanny. The Internet and World Wide Web, both important for the dissemination of Artificial Life research, are thickly grown over with organic metaphors. No longer symbols of bureaucratic rationality, computers, in the age of personal and networked computing, have become more "natural" to use. The *Whole Earth Review*, a magazine that in the wake of the 1960s extolled the virtues of getting back to the land through cooperative gardening, now advocates getting on the Internet as a way of rediscovering organic community. The modem has replaced the hoe as the technology of reunion with nature. In an age of environmentalism, it is no wonder that biodiversity appears online in Ray's global reserve for digital organisms. The cyberspace first envisioned in the science fiction of William Gibson has gone green.

Artificial Life technology is a kind of time machine signaling a quest to re-create nature in the image of a communication and control device devoted to information processing. Tierra offers us the opportunity to travel back to the dreamtime of the Cambrian period, back to an idyllic past full with possibility. Tierran organisms are not unlike the dinosaurs cinematically resurrected in *Jurassic Park*. Both are avatars of a digital nature produced through the magic of informatics. And this digital nature germinates in the circuits of more computers than those used in Artificial Life or in the Military Industrial Light and Magic Complex, Julian Bleecker's name for the closely related collection of Defense Department and Hollywood imaging technologies. The Human Genome Project, the multibillion-dollar transnational enterprise of mapping the full complement of human genes and of sequencing the billions of base pairs that compose human chromosomes, catapults human nature, written in the code of information, into databases that may increasingly define what counts as human being. The sequence map is often figured explicitly as a holy grail for biology (see Flower and Heath 1993), keying us into the themes of salvation history that inhabit informatics and molecular genetics at the turn of the second Christian millennium. As Artificial Life scientists remake the organism and replay the creation, they run their time machines forward as well as backward; like people who have their bodies frozen in the hope of future reanimation, Artificial Life researchers seek to wake their creations up in new but familiar worlds. When *Jurassic Park* daydreams are downloaded into computers, we often get resuscitations of Judeo-Christian creation stories, masculine self-birthing narratives, heterosexual family values, and utopian visions of a cyberspace capable of returning us to an Edenic world of perfect communication and commerce. In short, we get digitized nature programs about endangered fantasies. And we frequently get them with an environmentalist twist that reaches back to a 1960s countercultural sensibility. John Lennon, recently reborn as a computer-manipulated audio ghost, inspirits the newly reunited Beatles no less than he does the ecologically correct refrains of Tom Ray, one of a number of scientists who see Artificial Life as reviving countercultural promises to radically remake the way we see life and the universe. Artificial Life is a way of life in these digital days, and vital signs are routinely animated via computer. All of this doubling and déjà vu is oddly appropriate at this fin de siècle, as the year 2000 promises second chances and new and improved beginnings (see Schwartz 1996). The future has arrived just in time, and the fashioning of vitality as information processing—as the replication of code—has already brought us a kind of second coming in the flesh and blood of the lamb clone Dolly, who appears in the aftermath of nuclear DNA transfer, not nuclear war, and whose genes have been forced to reveal their essence as programs for self-duplication. Artificial Life joins cloning and cryonics in a trinity of millennial technologies of resurrection.

As organic logic is injected into computation, some of us may come to understand computers as effective because they operate using "natural" principles. At the same time, we may come to see the natural world embodying a computational calculus. The cultural commitments embedded in such visions may become increasingly difficult to discern as boundaries between natural and artificial process smear. It will be important to ferret out the ideas that nest within our constructions of nature, not so we can eliminate them, which would be impossible, but so that we can see that our pictures of nature represent a social accomplishment with which diverse, interested audiences can engage. Artificial Life is a project with a capacity for strewing reality with programs and machines modeled after already constituted social categories, programs and machines that might provide yet more "empirical" evidence for the idea that organisms really are automata, really do act selfishly, or really do need stable "sexual" identities to survive. This book is a critical political intervention into the reinvention of nature under way in Artificial Life; it is an argument against the digital naturalization of conventional visions of life and a petition for a greater sense of possibility in the Artificial Life world.

ARRIVING AT ARTIFICIAL LIFE

My path toward doing this study began in the late 1980s when I was in college at the University of California, Los Angeles (UCLA), studying anthropology and dabbling in computer science. Early on, I read Douglas Hofstadter's *Gödel, Escher, Bach* (1979), a book that wove a beautiful tapestry of analogy between mathematics, graphic art, music, computations, genetics, and language. I was enthralled by the possibility that computers and DNA might have something in common, that intelligent machines might be the offspring of human cultural evolution. In my sophomore year, I enrolled in a class in Artificial Intelligence and philosophy with Charles Taylor, a biologist who later became a prominent figure in Artificial Life. I remember how inspired I was by the final lecture in which Taylor argued that living organisms were nothing more than specially organized matter. Life, he said, was just a question of how matter was put together. Inanimate objects had something of life in them simply because they were made of the same stuff as we organisms. As an atheist, I found this idea eerie, and it awoke sensations in me that I could only consider numinous.

As I continued my studies, I fixed my imagination on science fiction–inspired dreams of studying the anthropology of future human evolution; I wanted to know how humans might evolve in outer space, how they might create intelligent computers, and how they might confront their creations as evolutionary rivals. I mined my boyhood fascination with *Star Trek*, *2001*, and *Close Encounters* for inspiration. I was committed to a strongly scientific view of the world, manufactured in part by interactions with my mother's father, a chemist and amateur astronomer whom I greatly admired.

My view of science changed when I began work as a biological anthropology tutor for UCLA's Affirmative Action Program. As a white person working with predominantly African-American and Latina/o students, I found myself in uncomfortable and odd positions as I tried to explain the instructor's interest in retrieving a scientific definition of race, when such endeavors had clearly been the cause of so much oppression. My students challenged me to think critically about my racially marked position as a messenger of a version of evolutionary biology—sociobiology—that offered genetic rationalizations for a system of social inequality in which people like myself were among the privileged. I had long felt myself to be anti-racist—since early experiences in integrated elementary schools—and hoped that science would be a tool I could use to bolster my belief in the irrationality of racism. I had not imagined that racism still inhabited the body of contemporary science. I was keenly aware of how scientific racism had been deployed in the past, since my German father had communicated to me the pain of knowing that his parents' generation had joined in the Nazis' ferocious project of racial extermination. My involvement with affirmative action also pushed me up against the politics of race in the university generally, especially as the Reagan-Bush administrations made ever-deeper attacks on civil rights, attacks that my co-workers and I felt as the university targeted our jobs for elimination. I applied to graduate school in anthropology with the idea of learning more about the social shaping of science.

I found out about Artificial Life when I arrived at Stanford University and took a course with the computer scientist Terry Winograd about the failure of Artificial Intelligence to produce truly intelligent machines. I came upon the proceedings of the first Artificial Life conference as I searched for fresh work on "nonrepresentational Artificial Intelligence" and was chagrined to discover sociobiologically inspired speculations about how to make computers into selfish, competitive, and xenophobic creatures. Soon after, I began work as a teaching assistant for courses in human evolution, an activity that amplified my concerns with science and race and that pressed me to reexamine my rather orthodox views on the relationship between biology, sex, gender, and sexuality. Though this tale glosses over uneven personal and professional transitions, it does mark out the steps I took toward studying Artificial Life. How I gained access to Artificial Life scientists is the next piece of the puzzle.

In 1991, I submitted a paper to the first European Conference on Artificial Life. The piece discussed how John von Neumann's theory of self-reproducing automata reoutfitted biblical stories of creation in cybernetic clothing (see Helmreich 1992). When the paper was accepted onto a panel on epistemological issues in Artificial Life, I traveled to Paris to deliver a corresponding talk. Afterward, I was approached by Steen Rasmussen, a physicist working on computer simulation in Santa Fe and Los Alamos. Rasmussen had studied philosophy in his native Denmark and had recently relocated to the United States to pursue interests in Artificial Life. Citing the German social theorist Jürgen Habermas, who has written about the role of communication in constituting a democratic public sphere, Rasmussen said he felt that a cultural and historical understanding was very important to facilitating informed public dialogue about Artificial Life. A series of emails between Rasmussen and me followed, and I obtained permission to do fieldwork at the Santa Fe Institute and to use the Institute as a point of departure into the networked community of Artificial Life. Over the summer of 1992, solidly from May 1993 to July 1994, and again in June 1995, May 1996, and June 1997, I conducted interviews, did archival research, attended lectures and conferences, learned a few simulation platforms in depth, and visited international research sites.[5] I was sustained primarily by Stanford, by a grant from the National Science Foundation, and by the hospitality and staff support of the Santa Fe Institute. The imagery of travel speaks to the privilege with which I was able to move among researchers; I was not an exile, a refugee, or a migrant. I was assimilated into the Artificial Life community with minimal difficulty.

The historian of science Donna Haraway has commented that we should not undertake studies of activities in which we are not implicated, in which we do not have a stake. Though I am not a practicing Artificial Life scientist, I share an investment in examining the assumptions of contemporary biology and computer science. In a world in which computers organize much of everyday life and in which informatic theories of biology guide understandings of human nature and culture, I care about how knowledge about the organic and the digital is produced. I continue to find this knowledge compelling and disturbing, persuasive and problematic, and full of pleasure and danger. I read the following statement of Langton's as an invitation to do my study:

The *practice* of science involves more than science. Although scientists often work within a world of abstraction and mathematics, the results of their abstract mathematical musings often have very tangible effects on the real world. The mastery of any new technology changes the world, and the mastery of a fundamental technology like the technology of life will necessarily change the world fundamentally. Because of the potentially enormous

impact that it will have on the future of humanity, on Earth and beyond, it
is extremely important that we involve the entire human community in the
pursuit of Artificial Life. (1992:20)

Langton's remarks, while ambitious in their vision of the influence of Artificial Life, open a door to considering the science from a cultural perspective. This is perhaps not so surprising, since Langton's undergraduate training was in anthropology, a fact that fundamentally informed my
conversations with him and that shaped his generally sympathetic attitude toward my project. Of course, as I will be arguing throughout this
book, I do not think a cultural analysis of Artificial Life can be restricted
to the impacts of the field, as Langton seems to suggest. Artificial Life is
already informed by culture, history, and politics, even unto the most
"abstract mathematical musings" that animate it. Recent literature in
the social and cultural study of science has brought such recognitions
into sharper relief, and I am inspired by this work here, a brief account
of which is in order.

SCIENCE AS CULTURE

Culturally dominant pictures of science portray it as a social practice
dedicated to producing objective knowledge about the world. This image has a variety of institutional and epistemological supports, not least
of which is a scientific education that instills scientists with a sense that
they are committed to truth and pursue it in a community organized
around professional skepticism. This is a picture that sociologists of science have been busy disturbing for the past twenty years or so. Following philosophers of science such as Thomas Kuhn (1962), who argued
that scientists' knowledge of the world is organized by paradigms—that
is, culturally agreed upon ways of looking at nature—rather than by un-
problematic access to nature, sociologists of science have examined how
scientists craft and contest knowledge in social context. On their view,
nature does not exclusively direct the production of scientific knowledge;
theories are never completely determined by data, and decisions about
what counts as nature are often quite social, even political.

Recent literature in the sociology of scientific practice has focused on
the laboratory customs of experimental scientists, seeing the production
of scientific knowledge as a contingent and agonistic contest between social actors to marshal institutional and rhetorical support for truth
claims (see particularly Knorr-Cetina 1981; Latour and Woolgar 1986;

Latour 1987). Such a view, however, has often left the social meaning of scientific knowledge itself unexamined and has neglected to discuss how scientists participate in a system of values, beliefs, and practices that extends beyond the walls of the lab. It has often made no difference to the authors of these works whether knowledge is being manufactured about a protein, a pulsar, or a primate, or whether this knowledge is produced in the service of government regulation, product development, or lay activism. New work in the anthropology and cultural study of science has begun to pry at the cultural meaning of scientific knowledge and has examined scientists as people located in a multiplicity of positions (see, e.g., Traweek 1988; Haraway 1989; Martin 1994; Gray 1995; Edwards 1996; Fujimura 1996; Gusterson 1996; Nader 1996; Rabinow 1996; Galison 1997; Downey and Dumit 1998). Much of this work has built on feminist and anti-racist critiques of science (see, e.g., Bleier 1984; Keller 1985; Harding 1986, 1991, 1993; Haraway 1991a, 1991b, 1991c, 1991d, 1991e; Boston Women's Health Book Collective 1992; Noble 1993). These critiques, historically linked to radical science movements of the 1930s and 1970s, have been grounded in political activism concerning issues such as women's health and the contestation of racist theories of intelligence.

The anthropology of science has begun to dismantle the notion that science is carried out in a world scissored off from the culture in which it exists. Haraway provides a programmatic summary of the view that science is culture in her monumental history of twentieth-century primatology, *Primate Visions:* "My argument is not that 'outside' influences have continued to determine primatology into the period of problem-oriented, quantitative studies, but that the boundary itself gives a misleading map of the field, leading to political commitments and beliefs about the sciences that I wish to contest" (1989:125). I would like to say the same of this study of Artificial Life, and assert that it is no surprise that scientists use "extrascientific" resources in putting together models. The use of resources "outside" science is not a scandal but is science as usual; that we have been encouraged to deny this is the scandal, and is precisely the way an educated public has been prevented from participating in building scientific views of the world. As the sociologist Pierre Bourdieu writes, "The idea of a neutral science is a fiction, an interested fiction which enables its authors to present a version of the dominant representation of the social world, neutralized and euphemized into a particularly misrecognizable and symbolically, therefore, particularly

effective form" (1975:36). As members of a powerful institution that derives much of its authority from claiming to transcend human culture,
scientists have attempted to enforce through their practice and language
the notion that science stands apart from human affairs. Science has become a sacred domain.

This is not to say that there is not something distinctive about scientific practices, about modes of thinking and acting made available in the
physical and life sciences. These affect domains apart from science—
economics, politics, religion—as much as the reverse. A relativist position, one that holds that all knowledges are equal, ignores the ways science produces powerfully persuasive accounts of what is real, about the
world we share and shape. By being forgetful about the conditions of
power that mold knowledge, relativism is as bad as objectivism, the position that knowledge can exist without knowers. A view that sees science
as social accepts that science has been crafted as a particular way of knowing even as it is crosscut by cultural practices and commitments. Understanding science as culture opens up space for reimagining and intervening
in its projects. As the anthropologist David Hess (1992) argues, science is
always embedded in a contestable symbolic order and set of power relations and is therefore subject to cultural critique, the object of which is to
show that taken-for-granted practices are often quite contingent and
could be other than they are (Marcus and Fischer 1986). On this view, Artificial Life does not extrapolate eternal rules of nature into a machine
realm but transcribes culturally particular tales into its new creations.[6]

MATERIALIZING NATURE, LIFE, AND CYBORGS

In spite of its novelty as an ethnographic object, Artificial Life is interesting for quite traditional anthropological reasons. The field promises
both to reinforce and to disturb the stability of "nature" and "culture,"
categories that have long been central for organizing Western folk and
scientific thought. Nature has meant for many of us that which is moral,
inevitable, given, perhaps rationally or harmoniously designed.[7] In many
cases it has not mattered whether we appeal to a nature made by God
or to nature as it is revealed by science; it has been a reference point for
understanding some things as immutable (see Yanagisako and Delaney
1995).[8] As such, it has frequently served as a resource for legitimating
social orders—for naturalizing power. As long as nature occupies this

politically potent place, it will be critical to examine the cultural tales imported into it.

But Artificial Life may well participate in changing some dominant meanings of nature. "Artificial Life" is, of course, a deliberate oxymoron, meant to shake us into considering the possibility that "life" might not be an exclusively "natural" object or process. If, as the sociologist Karin Knorr-Cetina (1981, 1983) holds, the nature that scientists work with in labs is already highly artificial, and can be seen as the outcome of histories of local, practical decisions about what works (rather than about what is true), then Artificial Life, with its reliance on constructed computer simulation, might make it clear that the only kind of nature we can have is a kind generated through what Haraway has called a "relentless artifactualism" (1991e:295). The anthropologist Paul Rabinow has asserted that in the culture of late modernity "nature will be known and remade through technique and will finally become artificial" (1992:241–242). Practices like Artificial Life may radically reconfigure what we think of as life, drawing attention to the fact that our understandings are constructions in the most literal sense, that they are built from our imaginative and technological resources (see also Emmeche 1991, 1992, 1994; Doyle 1997b).

In her work on the anthropology of new reproductive technologies, Strathern (1992b) examines the shifting relationship between the natural and the artificial in late-twentieth-century Western life. She observes that Europeans and Americans seem increasingly interested in making explicit what they mean by "nature." This happens when they intervene technologically to help "infertile couples" have their "own" "natural" children, when they participate in the conservation of rain forests, when they engage in genetic engineering to change properties of plants and animals, when they produce and consume "natural" foods, or when they program organic dynamics into computers. As they seek to define the natural, they redraw what counts as natural and cultural, solidifying these as categories of thought, but also making nature available for analysis as a culturally constructed item. The distinction between first and second nature erodes.

The idea that nature is what we make it is called "social constructionism." Social constructionism has been a politically powerful tool for contesting naturalizations of inequality. People interested in dismantling structures of domination organized according to categories of gender, race, and sexuality and through economic patterns like capitalism have been eager to show that these categories and patterns are not natural, not biologically given, but are the result of historical and cultural processes. In recent decades, sociobiology has been just one prominent focus of social constructionist debunking. While such critical deconstruction has been useful, it has begged the question of what nature remains when the work of social construction is done. Social constructionism has reproduced as residue the very nature it has sought to dethrone.[9]

Paradoxically, a social constructionist attitude fundamentally enables the science of Artificial Life. Researchers are explicitly concerned with creating new biologies *in silico,* an enterprise that forces them to question whether nature is reflected or constructed in their work. While I find such questioning epistemologically promising, I would like to offer an alternative to social construction, one I borrow from the rhetorician Judith Butler, who proposes that we think of the reality of categories like "sex," "race," and "nature" not as constructed but as *materialized* (1993:9). This means that the realness of things congeals from material practices of meaning making that stabilize "over time to produce the effect of boundary, fixity, and surface" (1993:9). An object or process like "life" does not exist "out there," waiting for us to name it. Neither is it solely the product of active "construction." What "life" is or becomes is "materialized"—comes to matter (in the sense of both becoming important and becoming embodied)—in such practices as describing and fabricating machines and organisms with common metaphors and taxonomies, negotiating boundaries and connections with nonhumans, representing living beings as ordered by their visible structure rather than their smell or taste, and so on. To borrow an argument from Haraway about nature, the fact that "life" exists for us at all "designates a kind of relationship, an achievement among many actors, not all of them human, not all of them organic, not all of them technological" (1991e:297). This book, though centered on the human agencies enlisted in the making of Artificial Life, tries to get at how new notions of life are being materialized, specifically, at how life is being crafted to inhabit both the natural and the artificial—a process that is already transforming our meanings of nature, evolution, and life. This mutation bears watching because, as Haraway notes, "if technoscience is, among other things, a practice of materializing refigurations of what counts as nature, a practice of turning tropes into worlds, then how we figure technoscience makes an immense difference" (1994:60).

The object that Artificial Life researchers are seeking frenetically to mime, to reproduce, namely "life," has been notoriously difficult to define. And as researchers forward new candidates for vitality, they both stabilize and undermine any definitional enterprise. The historian Michel Foucault reminds us that the category "life" is in fact a relatively recent invention:

> Historians want to write histories of biology in the eighteenth century; but they do not realize that biology did not exist then, and that the pattern of knowledge that has been familiar to us for a hundred and fifty years is not valid for a previous period. And that, if biology was unknown, there was a very simple reason for it: that life itself did not exist. All that existed was living beings, which were viewed through a grid of knowledge constituted by *natural history.* (1966:127–128)

Only with the rise of Darwinian notions of evolution and accompanying ideas about the underlying relatedness of all living things did it become possible to conceive of life as something in itself. "Life" emerged at the end of the nineteenth century as an invisible unity, a force or principle unifying the visible forms of living things.[10] "Life" became an elusive quality, a "secret" to be sought in the threads that tied living things together, an essence to be located in the filaments that maintained relationships between living beings. In the twentieth century, scientists came to see "life" as residing in the substance of DNA, in a "code" that could be read. This definition of life has materialized new life-forms, such as genetically engineered animals and plants, as well as new ways of thinking the human, as a cybernetic organism endowed with a genetic potential and profile. The extension of the genetic code metaphor to grant computer programs vitality promises unexpected new materializations of "life." Locating changing concepts of animation in cultural context matters because understanding and managing "life" has become an important political activity; one has only to think of the place of "life" in debates about environmentalism, abortion, euthanasia, and new reproductive technologies (and see Taylor, Halfon, and Edwards 1997). As a trend-setting science, Artificial Life may provide a window into the changing scientific definitions of life that crisscross these controversies.

Foucault argued that discourses that fence in life first became potent political items in the nineteenth century. He designated "bio-power" as that which "brought life and its mechanisms into the realm of explicit calculations and made knowledge-power an agent of transformation of human life" (1976:143). Control over definitions of life has made our bodily practices the subject of disciplining technologies and knowledges. Bio-power operates less through controlling people than by defining them, by constituting their identities such that they believe they fit into already extant natural categories. Bio-power populates reality with subjects constituted according to its definitions, thereby providing empirical evidence that the categories are real. It operates through religious, scientific, medical, educational, and state institutions, as well as through the politics of everyday life—in the bedroom, the kitchen, the office. It circumscribes the space of the acceptable, gives rise to new modes of subjectivity. If men and boys feel guilty for masturbating because they believe in some way that they are dissipating their "life-force" unproductively, this is an operation of bio-power. If women's menstruation is constructed as disqualifying them from "normal" political life (modeled after a male norm), then bio-power is at work here, too. As the organic and the technological merge, mix, melt, and mutate, however, we might be better served to look through the lens of the "cyborg," the human-machine hybrid first theorized in the mid-twentieth century in the service of the U. S. space program's mission to integrate humans with extraterrestrial life-support systems (Clynes and Kline 1960). Haraway may be right that "Michel Foucault's biopolitics is a flaccid premonition of cyborg politics" (1991d:150).

Seeing the cyborg as a prosthesis for cultural theorizing comes famously from Haraway (1991d), who proposed the cybernetic organism as a figure for a new politics of connection, affinity, and boundary transgression. As a feminist coming of age in the 1960s and 1970s, Haraway was wary of identifications of "women" with "nature" and desired attention to the way identities were socially built. Identifying women with nature may have been empowering for some, but it invited patriarchal appropriation; if women were closer to nature than men, they were leashed to whatever dominant biological science decided their natures were. The cyborg, continually changing in response to new feedback, not entirely natural or artificial, could be a more elusive and empowering figure, never stable, never fixed by original difference. In a sense, Haraway argued, we are already cyborgs: our bodies and identities are amalgams of the natural-technical codes of DNA, human language and its metaphors, the history of sexism, racism, and colonialism, and the ergonomic logics of early and advanced capitalism. Haraway argued that rather than bemoan this state of affairs, we would do well to appropriate it for liberatory ends.

Cyborg politics can force us to examine and take responsibility for the ways human enterprise and the organic world are ever more implicated in one another. Recent years have seen the growth of "cyborg anthropology," a practice that seeks to examine "ethnographically the boundaries between humans and machines and our visions of the differences that constitute these boundaries" (Downey, Dumit, and Williams 1995:342). Cyborg anthropology acts amid the contradictions of technoscience, looking for liberating and oppressive stories commingling in the science we produce. My account of Artificial Life joins in this project of mapping the disturbed social and political boundaries between human and machine.

ETHNOGRAPHY, POWER, AND WRITING

With few exceptions, the subjects of traditional anthropology have been people in positions of less power relative to the anthropologist, who has been privileged to travel, to take time to immerse herself or himself in "another culture," and to produce a text that her or his subjects may never read. Doing anthropology among powerful people is different. Many do not like to be studied and can easily prevent anthropological access to their lives, although "this power can also produce what an anthropologist who has studied . . . elite institutions has called 'the confidence of class'— a sense that one's position in the culture is assured and unassailable" (Lutz and Collins 1993:48). The powerful have means to protect their interests and to contest their portrayal by journalists and social scientists. I must be aware that my writing exists in the same academic and cultural contexts as the people I write about.

In anthropology, the activity of studying powerful people is called "studying up" (Nader 1974).[11] The reasons for studying up are myriad, and include investigating the sources of one's political indignation at the practices of power, a curiosity to see how influential knowledges are produced, and an impulse to make the customs of the powerful available for public scrutiny. The consequences of studying up can be diverse; one may lose funding, get sued, gain unexpected allies, or be conscripted into a new way of life. The dynamics are not necessarily quite that simple, either. Within communities there are differentials of power: the power I had with respect to people in the Artificial Life community varied, as I entered into alliances, disagreements, and confrontations with people differently positioned in the field. My participation in the community meant having opinions, making friends and foes, and caring about its theory and practice. This book cannot be a definitive account of the social world of Artificial Life. It is a partial story, informed by my own history, training, and interests—as any story told by a located human being, in any kind of circumstance, must be. Nevertheless, I believe I have captured important cultural dimensions of Artificial Life work. I hope that Artificial Life workers recognize something new of themselves in my account.

There is something peculiar about the academic and institutional configurations that have allowed anthropologists to walk into high-tech laboratories, something that makes scientists laugh when I tell them what I do. I think this laughter bespeaks an uneasy recognition of the coercive and colonial relations that produced anthropology historically as the study of nonmodern "others," but it also reveals an increasing sense—even among those most disposed to believe in objectivity—that everybody's closest beliefs may seem strange when viewed from another angle. An anthropological common sense has infiltrated Artificial Life researchers' view of themselves to such an extent that many deploy a gentle irony when asked to reflect on their practices and beliefs. They know that they appear to others as singular characters.

Which brings me to how they appear as characters here. It is conventional in ethnographies to give people pseudonyms or to otherwise protect identities. People often speak in confidence, and such devices of disguise can protect them from being identified by those who might use their words against them. Sometimes people need to be guarded from uninvited questioning by other community members or from emotional, financial, and physical embarrassment or threat. Nothing I report will put anyone in great jeopardy, but I am still careful with the words I gathered. I have checked transcripts with every person I interviewed and asked them about the preferred disposition of their words. Some people appear with their real names attached to interview data while others are given pseudonyms. Sometimes people named in one passage appear under pseudonyms (or even anonymously) in another. Real names include surnames; pseudonyms do not. Through much of this book I rely on researchers' published papers and public talks, and for these I always give proper attribution.

A SEQUENCE MAP

This book is structured by images of travel through worlds real and virtual. Chapter 1 surveys the geography of Santa Fe and the Santa Fe Institute, exploring the history of these spaces and the people associated with them. Chapter 2 moves into the worlds that materialize within computer simulations. Chapter 3 is an extended tour of several simulation systems, notably Tom Ray's Tierra and John Holland's Echo. Chapter 4 follows the narratives of Artificial Life out of the computer and into the lives of researchers, asking how practitioners use their work to reflect on questions of personal and cosmological meaning. Chapter 5 reaches into the worldwide web of Artificial Life, paying particular attention to European networks and investigating how diverse definitions of "life" circulate in an increasingly diverse transnational community. The coda closes with a report on Artificial Life in Japan and with meditations on the future of the field.

Because Artificial Life metamorphoses so rapidly, I cannot hope to provide an up-to-the-minute report on the latest turns. As I write this introduction, many of the ideas I discuss here are beginning to sediment into a general common sense, in science and elsewhere. Simulation technology has become less extraordinary, as has the notion that living beings are information-processing systems. Thus this ethnography can be read as an account of how some of us came to think and act the way we do. It is not surprising that Artificial Life has settled into a comfortable past. In its very conception, Artificial Life imagines that it is always in danger of being outmoded, superseded by the products of its own practice. This book races alongside that imagination, with a view toward understanding the shape of things to come in an age of silicon second nature.

ONE

Computing is still very much a frontier science.

GREGORY J. E. RAWLINS, *Slaves of the Machine*

Simulation in Santa Fe

SANTA FE IS a small city in northern New Mexico, and it unfolds at the base of the Sangre de Cristo mountains. A pedestrian-friendly knot of buildings sits in the downtown area, and most of the rather dispersed city fans out from there. During the time of my fieldwork, the Santa Fe Institute (SFI) rested three miles out of the center of town.[1] Each day I took the Santa Fe Trails city bus to a nearby hospital. From there, I traipsed across the low desert brush between office buildings of lawyers, doctors, and insurance companies to arrive at the unassuming faux adobe complex in which the Institute was housed. The Institute was the anchor point for my study of Artificial Life, the site from which I began my voyages into worlds social and virtual.

Traditional ethnographies often open with a description of the surroundings in which subjects live. This device has the purpose of grounding the ethnography, showing how peoples' lifeways are constrained and enabled by their environment. I mean to use this device self-consciously, to locate people who sometimes feel they belong to an aterritorial—even acultural—community. In what follows, I describe SFI as a physical location, as an interdisciplinary research center, and as a node in the international networks of Artificial Life. I describe the people who populate SFI, registering the Institute's ethnic, national, gender, racial, and class composition. Moving out from SFI, I situate scientists in the geography and history of Santa Fe and northern New Mexico, reaching back from here into the recent past to retrieve origin stories about both the Institute and Artificial Life. I finish with a portrait of an Artificial Life conference in Santa Fe. The journey has a kind of fractal structure, folding back on itself to show how people are connected to a variety of histories and places on many different scales. I begin now with a picture of the office space of SFI as it appeared during my fieldwork, in the hope that this will give gravity to the social worlds to which I tether the computational worlds of Artificial Life.

INSIDE THE SANTA FE INSTITUTE

SFI's main building has two floors.[2] On the first, there is a public reception area, a set of offices for administrators and staff, a photocopying nook, a kitchen, a pair of sexed rest rooms, and a small room for lectures and workshops. This conference room, equipped with a whiteboard, a television, a VCR, and an overhead projector, holds about thirty people comfortably. When I was at the Institute, it was often stuffed full. On the second floor are the offices and common areas where researchers work. In most places in the Institute cool white walls meet one another at right angles, but in the spaces connecting offices they confederate into octagons to accommodate diamond-shaped skylights and planters full of large-leafed plants. The Institute also occupies the first floors of two adjacent buildings. The publications office is located in one building, and the Institute's small library, filled with a spotty collection of computer science, physics, and mathematics texts, is wedged into the other. The collection is not impressive, since SFI does not care to compete with the libraries of large research institutions.

The reception area in the main building is the public face of the Institute. Its walls are a quiet pink, and the pillows on the two inviting off-white couches are a polite agreement among aquas, mints, and mauves. Above one couch hangs a Navajo rug, and on the cylindrical white end table shared by both couches rests a gigantic quartz crystal, a gift to the Institute from wealthy benefactors. In their rustic simplicity, the rug and crystal evoke both a southwestern and a scientific aesthetic. The rug conjures a romantic vision of the indigenous people of New Mexico whose customs and arts have been appropriated and homogenized in southwestern decor and art—or Santa Fe Style,[3] as it is usually called. The design of the rug also suggests the systematic generation of complexity from the iteration of simple rules; it is possible to imagine weaving such an item using elementary computational procedures. The quartz crystal on the white pedestal stands as a New Age vision of nature's deep wisdom and potentially redemptive power. But its latticed structure also speaks of the discipline and beauty of physics. Throughout the Institute, scientific and southwestern aesthetics alternately bleed into and stand opposed to one another.

The logo of SFI, which adorns all Institute publications and is displayed on a downstairs wall, is a circular figure based on an ancient Mimbres pottery design (ca. 950–1150 C.E.).[4] This logo is meant to locate SFI as an institution in the southwestern United States, though it references cultural traditions that are not part of the heritage of most of the people of Euro-American descent and identity working at SFI. This referencing of indigenous American cultural forms is common at the Institute. The offices of administrators are decorated with prints of Native American pottery and dolls. On the stairwell that leads upstairs, there is a photograph of Albert Einstein wearing a crumpled suit and a feathered headdress (likely from the Plains peoples), posing with his wife among a number of Hopis (see fig. 1). What makes this photograph appealing and amusing to visitors—and it almost always gets a laugh—is the juxtaposition of Einstein, a symbol of twentieth-century science, with "primitive" Native American clothing and people (posed, we might note, in front of a simulated Hopi dwelling designed in 1904 as a tourist attraction by the Santa Fe Railway [Wilson 1997:113–114]). Einstein's headdress makes him look like a little kid playing Indian. I realized as I saw people respond to this picture that the Hopis, as far as most were concerned, were just part of the landscape.[5] The Hopi faces and what might be read from them were not interesting, a fact that pinpointed the gaze of many of the scientists as a white Euro-American gaze, one that did not see the picture speaking about the relations of inequality, exploitation, genocide, and tourism in which Native Americans and Euro-Americans have existed.

Upstairs are the sites where researchers do much of the work of computer simulation. Sofas, bookshelves, and computers of various breeds fill the common areas, serving as conversation pieces and places, as distractions and inspirations. Many outsiders find it striking that the only scientific tools in evidence at SFI are computers; there are none of the fixtures usually associated with sites of scientific work, no test tubes, no machines for amplifying DNA, no special sinks. The computers that sit in common areas and in offices are mostly Sun Sparcstations and, continuing the southwestern theme of SFI, are individually named after local tribes or pueblos, like Pojoaque, Picuris, and Nambe. The naming of workstations after Native Americans is reminiscent of the white

Figure 1.   Dr. and Mrs. Albert Einstein at Hopi House, Grand Canyon,
Arizona, February 28, 1931. Photo by El Tovar Studio. Courtesy Museum of
New Mexico, Negative no. 38193.

Euro-American practice of naming sports teams or cars after native
groups, a practice connected to the commodified romanticization of
these groups in popular U.S. culture (see Churchill 1994).[6] Remarking
on this is not to argue that SFI is acting in bad faith but to recognize that
a white imagination circulates in the Institute and—as I will argue
later—some of its associated science.

   Amid the computers there are a couple of secretarial desks crunched
into the corners of open areas. The walls of the second floor are covered
with framed artwork, posters, and bulletin boards. Geometric art and fan-
tastic photographs display mathematical principles and busily call attention to their techniques of creation. At one time, there hung a poster labeled "UNIX FEUDS" that showed a white wizard protecting a
light-skinned boy wearing a headband labeled "user" from a horde of bel-
ligerent and darker-skinned medieval soldiers fighting under the banners
of competing computer companies. Bulletin boards display calls for papers, conference announcements, maps of computer networks, and news-

paper cartoons about the quotidian activities of scientists. Coffee tables are arrayed with a scattering of computer science and physics journals, issues of *Science* and *Nature*, treatises by famous scientists, environmentalist publications, manuals, and the occasional book on virtual reality and computer games. Blackboards and whiteboards are available so that people can spontaneously jot down proofs, equations, models, and speculations. There are a few coffee machines located around SFI, and people use this ritual drug to give themselves energy for a productive workday.

Because one of the missions of SFI is to foster interdisciplinary conversation and collaboration, people usually share offices with researchers outside their field. This interdisciplinary mixing is complemented by intergenerational mixing; graduate students in their twenties routinely brush up against seventysomething Nobel Laureates. Most people leave their office doors open, both to invite conversation and to display their collections of posters, computers, and bric-a-brac. More publicly known people have begun to receive letters from bizarre admirers; particularly "kooky" epistles are often tacked to office doors.

If the interdisciplinary philosophy of SFI is mapped onto an open, socially centripetal architectural space, then the temporariness of most people's time at the Institute is indexed by the hastily laser-printed names affixed to office doors. The flow of people through the Institute generates a quiet confusion and a generally easy attitude toward personal space and property. There seems to be a communitarian ethic about sharing at least some computational resources; people frequently log onto others' machines to use different software, connect to the Internet, and take advantage of the different capabilities of Macintoshes, Sun Sparcstations, and NeXTs.

But though some people are gregarious, others are so absorbed in their personal projects that it is almost as though they are not really present—and perhaps in some sense they are not. Understanding SFI solely as a physical place is to ignore that for many it is an entry point to the computer networks that tie together their real intellectual communities. The Internet functions as a kind of *conscience collective* that bands virtual communities together across geographic space. Describing the set of people doing Artificial Life may begin in a physical place but must also record how the community is threaded together through the geography of computer networks.

That part of the social landscape of the Artificial Life community is situated in international cyberspace does not mean people are not positioned in real spaces structured by complex social relations. Almost all the people at SFI—staff and scientists alike—are white, middle-class, and of Euro-American or European descent (there were two Hispana secretaries at the time of my work). Prevalent U.S. gender inequalities line up fairly neatly with the conceptually and proxemically distinct categories of staff and scientists. Most staff are women and work downstairs.[7] Most scientists are men and work upstairs (about 90 percent of researchers are men, based on informal counts I did repeatedly at conferences, at workshops, and on "In Residence" rosters of SFI). Photographs of scientists are displayed on a bulletin board near the top of the stairs. Each photograph is coupled with an outline of the researcher's interests, leashed to a pin stuck into a world map. A glance at the map reveals a few things. It is a Mercator projection, placing the countries of the north on top and privileging the shapes and sizes of northern hemisphere continents. The distribution of pins on the map reveals that most researchers are from Europe and North America, and while we should be wary of conflating cultural background with geographic provenance (see Gupta and Ferguson 1992; Malkki 1992), the clustering does tell us something about the political economy of science. The map is meant to demonstrate how international SFI is; ironically, it suggests how culturally particular is the science done at SFI, how it depends on economies of privilege, education, and travel that connect elite cultures in Europe and the United States.

SFI is a bustling place, and a variety of scientists constantly come through, visiting for the day, attending a workshop or conference, engaging in a monthlong collaboration, or settling in for a year of residence. SFI is well known for its small size and attracts economists, computer scientists, biologists, physicists, mathematicians, and archaeologists from a variety of universities and research institutes. These people meet in interdisciplinary conferences on topics such as computer modeling of nonequilibrium ecological, economic, and technological systems; cultural evolution; and multiagent simulation systems. At any one time there are about forty scientists at SFI, and several are busy with thoughts about Artificial Life.

The age of scientists at SFI ranges from about nineteen to seventy. At the low end of the spectrum are undergraduate interns. Four or five pass through SFI each summer, working under graduate students or postdoctoral fellows on small, well-defined projects. SFI also hosts a yearly Complex Systems Summer School, which about sixty graduate and postdoctoral students attend—though few actually spend time at SFI, since instruction is held off-campus. Graduate fellows, of whom there are about four or five at any time, are affiliated with a regular university but are in residence for a year or two doing research hard to carry out at their home institutions. During my stay at SFI, I was placed in this category. Postdoctoral fellows, ranging in age from midtwenties to midthirties, are about as numerous as graduate fellows and stay at SFI for about three years.

The bulk of people at the Institute are between thirty and fifty, and most work as professors at universities around the United States and Europe and visit SFI for a year, a few months, or just a week. The oldest set of people are scientists retired from jobs at Los Alamos or other prestigious research locations. SFI was founded by some of these folks, and they maintain an elder statesman presence at the Institute, offering grand visions of the sciences of complexity and participating in and organizing workshops on topics such as sustainable development. While they are admired for past accomplishments, most younger people at the Institute prefer to collaborate with people in the prime of their research life.

All these categories are crosscut by gender. Fully half the undergraduate interns are women, somewhat fewer than that among the graduate and postdoctoral fellows and summer school students and fewer still among the set of itinerant and semipermanent faculty. There are no women among the eldest age set of scientists. SFI has expressed its commitment to equal opportunity, and through programs like the summer school, it is making efforts to encourage women to build scientific careers. But SFI is still located in a cultural context that has not encouraged women to enter science, and the historical legacy of this context tells on the distribution of women over different age groups. That women are much more powerfully represented among support staff led one staff woman I interviewed to say that she felt like she was working for all those smart boys she remembered from high school chemistry class. Another said that she was stunned the first time a woman was introduced to her as a conference participant. One woman scientist told me that when she answered the telephone after hours one night, the male scientist on the other end assumed she was a secretary, an assumption that greatly angered her.

Such incidents are rare, and the Institute nurtures a generally friendly atmosphere. In office memos and the *Bulletin of the Santa Fe Institute,* the complement of people at the Institute is routinely called "The SFI family." This usage calls on the relations of solidarity that many Americans feel characterize families—though it also papers over the hierarchies that often structure such groups (see Schneider 1968; Collier and Yanagisako 1987). Staff are invited to sit in on scientific workshops, though their office tasks often make this impractical. Many staff listen in on scientists' conversations and are, in various mixtures, skeptical, uncomprehending, and enamored of SFI science.

Because scientists are at a variety of stages in their careers, their economic positions are various; graduate students struggle to get by while middle-aged and older scientists buy land in Santa Fe and build houses. It is easy to discern the middle-class trajectory and aspirations of SFI researchers. People want to have their own house, car, time and money for vacations, perhaps a place to raise children. Most older researchers who come through the Institute are fairly well-off, and those who come for a short time rent cars and stay in nice hotels (about $100 a night). The *Restaurant Guide* provided to SFI visitors lists moderately priced places but also includes quite expensive restaurants (about $60 a person). I often heard older people talking about the Santa Fe art scene, and the names of well-respected and pricey artists like Wyeth and Miró were mentioned.

Most scientists dress casually, in jeans or shorts, T-shirts or button-down shirts. Many look as though they are ready for a short nature hike or jog, and some wear high-tech hiking boots suggesting a typically American penchant for rejoining nature through the use of sleek technology (see Nye 1994). Some men have adopted Santa Fe Style dress and sport bolo ties and turquoise accessories. Because SFI tries to maintain a relaxed feel, women scientists, who might feel compelled in other situations to dress in business attire, feel comfortable wearing casual clothes. At the same time, as several told me, they are careful not to wear too much makeup, and most favor pants over skirts. The women staff wear more stereotypically feminine professional clothing, including heels, silky blouses, and linen blazers. During my work, the few male administrators wore starched dress shirts, ties, sweaters, and slacks.

Another set of people gather around the Institute: the contracted gardeners, cleaning people, mail carriers, and trash collectors. These people are not considered part of the SFI family. They are mostly working-class and of Mexican descent. When gardeners are present on the periphery of outdoor functions, they are ignored by researchers.

My own movement as an anthropologist into SFI was facilitated by the fact that I fit well into existing categories and dispositions. My pres-

ence as a long-haired Stanford-affiliated white male in his midtwenties
was easy to assimilate, especially when people did not know I was an
anthropologist, which not everyone did, simply because there were too
many people coming through. At many moments, I had to make it a
point to introduce myself as an anthropologist, something that often got
a response not too far off from what happens when I do this in everyday life: people ask what geographic area I study or say something about
bones, archaeology, or nonhuman primates. Once researchers understood what I was doing, they often joked that Artificial Life scientists
were indeed an odd "tribe," a description that, while playful, imagines
the objects of anthropological inquiry inhabiting bounded spaces, possessing consistent mythologies, and leading lives distinct from others'.
Though the Artificial Life community does have a certain coherence, it
is also located in a net of larger and smaller cultural groupings and realities. Isolated tribes are largely an anthropological fiction, and I do not
care to perpetuate it here (see Wolf 1982).[8] That said, my first steps toward locating Institute Artificial Life scientists in a wider world require
us to take a trip through the city of Santa Fe.

OUTSIDE THE SANTA FE INSTITUTE

Santa Fe, capital of New Mexico, is home to about 63,000 people and
inherits a host of different histories, including those of the Pueblo and
Athabaskan (Navajo and Apache) peoples, the Spanish conquistadors
and their descendants, the Mexicans, and the U.S. settlers who partic-
ipated in projects of Western "frontier" expansion. Santa Fe was named
by Spanish colonists and means "Holy Faith." The city is in a spectacular setting. Seven thousand feet up, it lies in a bracing high desert valley,
saddled between the Sangre de Cristo and Jemez mountains, which are
spotted with piñon and juniper. It is pleasantly warm in summer and
brightly snowy in winter. What one notices most on arriving in the city
is that almost all the buildings are made to look as though they are constructed of adobe (a building material consisting of sun-dried earth and
straw)—that is, are made to conform to an image of Pueblo-colonial
Spanish architecture. But the present look of the city is not from mere
preservation of existing buildings. Santa Fe has actually seen many attempts to eliminate adobe construction. In the years following New
Mexico's 1912 statehood, however, architects like John Meem revived
and elaborated the style, with the result that the architecture of Santa Fe

references a prior reality that never really existed in quite the way it has been refabricated (see Wilson 1997). Today, according to the *Santa Fe, New Mexico, Fact Sheet, 1992–93,* "preservation of historic buildings is widespread and strict zoning codes mandate the City's distinctive pueblo and territorial styles of architecture." Santa Fe is in many ways a perfect place for a science based on simulation, a science in which practitioners try to make their simulations become real.

Close to 50 percent of the people in Santa Fe are of Spanish or Mexican descent. These people are often referred to as "traditional" or "nativo" by themselves and others and are the stable reference point people contrast to "outsiders" and "Anglos." A great many call themselves "Hispano" to emphasize lineal ties to Spanish families who owned a lot of land during the time Santa Fe was part of New Spain (see Nostrand 1992). Others use "Mexican-American," "Hispanic," "Latino," or "Chicano" for the various political purposes these signifiers have. All these groups are being marginalized as land prices and property taxes are driven up by well-off outsiders who consider Santa Fe a fashionable place to live. As wealthy whites "improve" their property (often, ironically, by making it conform to mythic Pueblo-Spanish architectural style), it is becoming impossible for many "traditional" people to stay on land that may have been in the family for years. Older people are losing their land, and younger ones are losing hope of buying land. Many are forced to move out of the historic center of town into the far southern outskirts, a place of strip malls, fast food, and trailer parks. There is an explicit racial element to all of this, and some longtime residents have complained of a "white invasion." The economic and racialized inequalities of Santa Fe have contributed to a city that suffers strong de facto segregation and territorialization. There are efforts to solidify community centers and churches in Hispano neighborhoods, but these communities are under siege by real estate developers, by unfavorable city planning, and by white newcomers who are buying up land and transforming portions of old neighborhoods into exclusive gated communities.

Most SFI researchers do not live in Hispano neighborhoods, in part because there are few rentals in these areas, but also because the Institute often finds housing for visitors. During my summer of preliminary fieldwork, I was placed in a well-ordered neighborhood in the process of gentrification. When I returned to Santa Fe to begin my year there, I asked where I might go for listings and was referred to upscale health food stores and coffee shops. SFI has an enormous effect on how scientists are placed in the city. The Institute encourages scientists to traverse the same ethnic space as white tourists; in fact, the tourist guide to Santa Fe, which is written for an audience of suburban U.S. families, is distributed to scientists when they arrive. A walking map of Santa Fe's "Historic Guadalupe District," centered around the oldest shrine to the Virgin of Guadalupe in the United States, explicitly does not include the adjacent predominantly Spanish, Mexican, and Mexican-American neighborhood.

Only 2 percent of people living in Santa Fe are Native American. Most Native Americans live out of town, on reservations or in other nearby settlements. Those who live in town are mostly Pueblo and Navajo. The Native American presence in Santa Fe manifests itself mostly for non–Native Americans through the annual Indian Market, the vendors who sit in front of the Palace of the Governors on the city's central plaza, the Institute of American Indian Arts Museum downtown, the nearby pueblos visited as tourist sites, and the Native American art and designs that decorate the buildings and fill the stores of Santa Fe. In recent years, some native groups have protested against their representation in Santa Fe tourist culture, making activist art and withdrawing participation from the city's annual Fiesta, an event that often celebrates Hispano culture without recognizing its roots in Spanish colonialism (see DeBouzek and Reyna 1992).

Although Native American motifs figure greatly in the Institute's image, no Native Americans work at SFI. Many scientists who come through Santa Fe visit the Indian Market, and some make a point of buying "authentic" crafts or attending local powwows at which they can experience Native American culture as a kind of edifying entertainment. For one SFI gathering in downtown Santa Fe, held at the Palace of the Governors, a site that has acted as the seat of Spanish, Pueblo, Mexican, and U.S. rulers, the Institute hired Albert Cata, a Pueblo-Seminole, to tell traditional stories to scientists and their guests. SFI sponsors occasional field trips to Native American archaeological sites, some of which are managed by Native Americans, like the Puye Cliffs in Santa Clara Pueblo, administered by the Tewa.

There is a thriving gay, lesbian, and bisexual community in Santa Fe, mostly white and affluent. Each year there is a substantial pride march, and there are a few openly gay-owned businesses in the city. Though there are and have been a number of gay people at SFI, most choose to remain closeted or silent about their sexuality. One scientist passing through the Institute told me that he wore a pink triangle for a while but soon felt uncomfortable and sensed that this was not an appropriate venue for a statement of this "private" aspect of his identity. Heterosexual identity is of course quite public, as I was reminded again and again at SFI by email messages that announced the births of babies to straight couples.

About half the population of Santa Fe is white, primarily of Euro-American descent. Local parlance refers to these people as "Anglos," a term that is also "used colloquially to refer to anyone who is neither Hispanic nor American Indian, even if that person does not have a northern European heritage" (Hazen-Hammond 1988:1). This group includes ranching families that have been in the area for generations. It also includes folks who arrived early to midcentury as anthropologists, and whose kindred descendants gravitate to the School of American Research, a famous anthropological research center. Santa Fe has also become a mecca for artists, and is the second-largest art market in the United States (after New York City), a fact tied to its status as a place of unconventionality and style and ultimately to the fame of the New Mexico artists Georgia O'Keeffe and R. C. Gorman.

Beginning in the 1960s and 1970s, a new set of mostly white people began trickling in who saw Santa Fe as a spiritual center. People arrived who were interested in "getting back to the land" and in exploring alternative lifestyles, religions, and healing. Some studied Buddhism, Taoism, and Native American religion. These folks have become the foundation of what might loosely be called a New Age community, which from modest beginnings has become rather large, even quite an industry (see Brown 1997 for an anthropological account of New Age culture in Santa Fe and elsewhere). During the time I was in New Mexico, the Whole Life Expo, "the world's largest exposition for holistic health and new age awareness," was held in Santa Fe (there are ten to twenty such events every year, in different cities). It was filled with exhibitions on astrology, alternative medicine, ecological home design, and crystals, as well as lectures on UFOs, tantric sex, channeling, past lives, and how to achieve physical immortality. Talks on how to use psychic power for financial gain indexed the way much New Age practice revolves around self-help rhetoric, a language that embeds particularly American ideals of success through individual effort. This language offers optimism to the largely white-collar baby boomer constituency of the New Age in Santa Fe, a group brought up to expect fiscal stability but who now find themselves forced to move and change jobs a good deal.

Most scientists at SFI are wary of any association with New Age movements, but because SFI tries to sell itself as doing cutting-edge, ecologically and socially concerned science, even, sometimes, "fringe" science (during my fieldwork an article on SFI appeared in *Omni,* which regularly takes UFO claims seriously), many of its formulations find resonance in the New Age community, something Institute scientists are continually on guard against. The social theorist Andrew Ross has argued that New Age culture exists in inevitable complicity with rationalist science even as it attempts to dispute its claims. It is "a social movement founded on an *alternative* scientific culture, distinct from dominant values, that is increasingly obliged to wage *oppositional* claims lucidly obedient to the language and terms set by the legitimate culture" (Ross 1991:532; see also Hess 1993). New Agers are constantly appropriating the language of technoscience to speak of how individuals might reprogram themselves, how people might take control of and guide the future evolution of the human species, and how virtual reality might be used to usher in an era of electronic shamanism. Ross writes that "New Age rationality . . . can be seen as a countercultural formation in an age of technocratic crisis. This crisis appears at a time when the official legitimacy accorded to technology-worship has guaranteed it the status of a new civil religion in North America" (1991:533). SFI scientists are aware of how science has sometimes acted as a force of destruction—it's part of the reason they tackle problems like sustainability, with an idea that this might contribute to kinder, gentler science. A few Artificial Life researchers share interest in alternative healing and thinking, and some embrace the chance to go to Zen retreats, to practice Buddhism, to do Sufi dancing, or to talk freely about experiences with psychedelic drugs. Often, they consider these to be private interests separate from their public lives as scientists.

The disdain with which many scientists regard the New Age community masks some of the links between alternative spirituality communities and SFI.[9] Scientists and New Agers share common class and ethnicity positions, though most of the scientists are male, and there is a skew toward more women in the alternative healing communities. The New Age community, like the SFI family, fixates on lessons we might learn from "nature"—often understood to be a self-regulating system with a certain wisdom. Both New Agers and SFIers share a romantic view of

Native American cultures and imagine them as well bounded, pristine, closer to nature—an image that has been produced by anthropologists of the Southwest, among others. And both New Agers and Artificial Life scientists refer to the potential existence of extraterrestrials to motivate their interests in theorizing alternative forms of life. It is not surprising that Santa Fe has been home not just to SFI but also to the Space Sciences Center, a nationally famous UFO museum.

There is a set of more recent white arrivals to Santa Fe: the relatively well-off people who come here to retire and participate in Santa Fe Style. Santa Fe's downtown is overrun by high-priced boutiques and galleries selling southwestern potteries, reproductions of Native American clothing, Kachina dolls, cowboy hats and expensive leather goods, and romantic paintings and sculptures of generic Native Americans. The growth of this well-off community and the tourist industry has created a spectrum of low-paying service jobs, occupied predominantly by local Hispanos, Mexicans, and whites who work at sequences of temporary jobs well below their educational and experience levels. A few older SFI researchers socialize with the wealthy new arrivals, and some have close connections to real estate dealers in town.

I have left out an important thread in the history of white migration into Santa Fe, and this is the line that follows the movement of scientists into the city. Beginning in the 1940s and 1950s, with the Manhattan Project and its spin-offs, scientists from Los Alamos National Laboratories started to move into Santa Fe. As nuclear weapons research became a permanent feature of the northern New Mexican economy, many people arrived from metropolitan centers. For those bound for weapons work and physics research in Los Alamos, Santa Fe became a haven from the militarized and business-oriented world of the lab town, which was crosshatched by security fences and checkered with unremarkable prefab homes and offices. The relatively high salaries of physicists at Los Alamos allowed many to purchase homes in Santa Fe, just under an hour's drive south.

For many white arrivals in New Mexico, Santa Fe represents the Old West, the "American frontier," and unspoiled nature.[10] One Artificial Life scientist in his early forties complained to me about the disappearance of nature in the modern era and said that he chose to live in New Mexico because there were still some bits of nature "unsullied by man." From his yard he said he could see the moon and the Milky Way, and this made him feel connected to something "primitive." Another said that it may be that SFI scientists who are interested in studying nature are interested in coming to "a place where there is a lot of it." If SFI were in Detroit, he said, it wouldn't have worked. SFI literature itself has been explicit about the role of "nature" in drawing people to the place, as a snippet from a brochure about the new SFI campus suggests:

> SFI's visiting scientists have also expressed excitement that they will be in touch with the restorative powers of Santa Fe's natural beauty. The facility is located in the middle of hills covered with piñon and juniper. For those who need the occasional walk to stimulate thought, or to clear the brain, nature awaits just outside the door. Others will step outside to sit under the wide portals surrounding the south and west of the house and let the view over the Rio Grande Valley and Santa Fe work similar rejuvenation. (Santa Fe Institute 1994a)

Nature and its life-giving (even rebirthing) power are highlighted here, echoing a sense that the projects of SFI are to be life-affirming. SFI is sometimes considered the good twin of Los Alamos, concerned with the technology of life rather than the technology of death. The nature imagined by SFI, stationed on a pure and untainted frontier, is a vision most characteristic of urban and suburban white Americans, for whom nature reserves and national parks are manufactured as pristine memorials to an "original" nature, a nature empty of politics and other people (i.e., Native Americans). The appeal of Santa Fe to scientists from Los Alamos is a crucial part of the story of the Institute, and it is worth traveling up the road to the lab to see how SFI was founded and how Artificial Life emerged from projects under way at Los Alamos in the late 1980s.

NORTHERN NEW MEXICO ORIGIN STORIES

SFI was founded in 1984 by scientists affiliated with Los Alamos. George Cowan, Murray Gell-Mann, David Pines, and Phil Anderson were theoretical physicists reaching a time in their careers when they could entertain the idea of founding an interdisciplinary research center. The credit for the initial idea usually goes to Cowan. Using their prestige (Gell-Mann and Anderson both received the Nobel Prize in physics), these men enrolled a number of others, including the Los Alamos senior fellows (a group of older scientists released from administrative duties). All these men were well connected in academia and, because of their association with Los Alamos, also had contacts and friends in funding agencies and the government. Cowan had been head of research at Los Alamos and had served on the White House Science Council.

The idea behind the Institute was to foster interdisciplinary research in what were coming to be called "the sciences of complexity." These scientists were interested in applying their thinking to grand problems in neuroscience, economics, evolution, and the information sciences. They had a faith that brains, economies, organisms, ecologies, and societies were all entities that could be considered "complex systems." Such systems were characterized by the fact that they were made of simple interacting elements that produced through their aggregate behavior a global emergent order unpredictable simply through analysis of low-level interactions. Brains were composed of neurons, economies of individuals and companies, ecologies of interacting species, and in none of these systems were global patterns predictable from constituent components. These systems were also called nonlinear, since relationships between variables were not straightforward and could not be described with linear equations. The scientists believed that the computer would be a perfect tool with which to study such systems, since nonlinear equations are difficult to solve by hand and computer modeling allows one to simulate consequences of nonlinear interaction without fully solving equations. Some felt that computers were themselves complex systems and that this made them perfect objects of study in their own right. The hope that complex adaptive systems from ecologies to economies might have something in common was fueled by the existing knowledge that many nonlinear problems in different fields, such as fluid dynamics and thermodynamics, had been described using cognate mathematical structures. The advent of computers capable of running an enormous number of calculations simultaneously and of churning through thousands of iterations very quickly also made feasible a science based on simulation. As the institutional voice of SFI put it in one annual report, "Modeling can provide approximations of solutions to problems that are otherwise intractable or can offer insights by simulating the behavior of systems for which a solution is not a well-defined concept" (Santa Fe Institute 1994b:4).

Although Los Alamos already supported a set of people working in nonlinear, chaos, and complex systems research, the founders of SFI felt that a new research center could provide an open, unclassified, and truly interdisciplinary site. It could orbit around Los Alamos, using some of Los Alamos's computational and personnel resources, but exist for itself. It would not be cannibalized into an existing university, since it would be too far from the University of New Mexico (UNM) in Albuquerque and certainly too far from major research centers on the East and West coasts. One person I interviewed said that SFI occupied a Legrangian point between Los Alamos and UNM, far enough from both not to get pulled into the gravitational field of either. The presence of Los Alamos has been crucial in getting SFI off the ground, and SFI still shares researchers and resources with the lab. People told me that researchers enjoy being at SFI more than Los Alamos since there are no security clearances required and it is not necessary to work on projects attached to military interests. And one can always telecommute to Los Alamos from picturesque Santa Fe.

The broad intellectual charter for SFI, which was to have an important effect on how Artificial Life settled into the Institute, was framed in 1984 when Murray Gell-Mann spoke at SFI's founding workshops. Gell-Mann outlined what he saw as the emergence of new syntheses in science. He contended that since World War II the physical, life, and social sciences had been drawing closer in a "striking phenomenon of convergence in science and scholarship" (1987:1). He credited the use of computers as theoretical and experimental tools with fostering cross-disciplinary conversations and with revealing deep similarities across the object domains of the sciences. One of these similarities he tagged as the idea of "surface complexity arising out of deep simplicity" (1987:3). Gell-Mann unabashedly designated himself a reductionist and made his belief in a connected unity of science evident by suggesting that all scientific explanations could be harmoniously integrated and then melted down en masse into the unifying pot of particle physics.[11]

In this first gathering around the concept of the Santa Fe Institute, held at the School of American Research, before SFI had quarters, Gell-Mann spelled out what a typical SFI conference might look like:

> Conferees will listen to reports on game theory strategies in biological evolution, the coevolution of genotype and phenotype in biological evolution, theoretical and experimental results on chemical or prebiotic evolution, the development of foraging strategies in ant colonies, strategies for the evolution of new algorithms in artificial intelligence (using crossing-over and natural selection in computer programs), models of human learning, the mathematical theory of regeneration in the visual cortex, discoveries on cellular automata and Turing machines, stability of deterrence and the U.S.-U.S.S.R. arms competition, spin-glass models of neural networks, and other diverse topics. Yet the discussion is to be general, with physicists, mathematicians,

population biologists, neurophysiologists, social scientists, computer scientists, and engineers all trading questions and comments. (1987:6)

There are many precedents for trying to unify the sciences,[12] and many depend on physicalist reductionism, but what makes the effort of SFI substantially new is the use of computers as tools for facilitating dialogue across disciplines and for committing researchers to a view of their different domains as permutations of the same underlying informational processes.

While the notion of the Santa Fe Institute was being consolidated by older men like Gell-Mann, there were a few people, also at Los Alamos, of a different age set hatching ideas that were to be the beginnings of the venture of Artificial Life. J. Doyne Farmer, Norman Packard, Chris Langton, and Steen Rasmussen were among a set of physicists and computer scientists in their late thirties who were becoming interested in how computers might be used to study biology. In September 1987, the first conference on Artificial Life took place at Los Alamos, cosponsored by Los Alamos's Center for Non-Linear Studies, the newly formed Santa Fe Institute, and Apple Computer. Here again among the core group was a set of mostly physicists fixing their attention on exciting new problems. And here again was a group of confident men forming a kind of brotherhood around a new and exciting concept, a brotherhood of men all socialized to take risks in life and career. The masculinity of the tale is highlighted by various popular science books about Artificial Life and SFI, in which we are introduced to key figures described as "maverick"[13] heroes questing for new knowledge. This vision was reinforced by tales told to me about how all-male camping trips and skinny dipping helped some members in this group bond. In my interviews, I discovered that early incarnations of the Artificial Life group often included the computer scientist Stephanie Forrest and the neuroscientist Valerie Gremillion, a fact that is frequently erased in popular accounts. A 1992 issue of the *Whole Earth Review* devoted to Artificial Life contained interviews with the above-mentioned men (along with some others), leaving figures like Forrest and Gremillion aside. Both have continued work in Artificial Life, however, with Forrest going on to be prominent in the area of adaptive computation.

While the Artificial Life fold soon expanded to include many computer scientists and became articulated to a set of projects that had been under way internationally in theoretical biology, Artificial Intelligence, and robotics, it is important to understand the place of Los Alamos theoretical physicists in the consolidation of these projects under the catchy moniker "Artificial Life." This shapes the ambition of the project, its focus on understanding biology as located in a world governed by the physics of information, its domination by men (theoretical physics is almost entirely male), and its claims to be a biology that will be true anywhere in the universe, just as physics is true everywhere in the universe. After the Los Alamos conference, Artificial Life, in its New Mexico incarnation, became ever more closely associated with SFI. Artificial Life scientists described biological systems as nonlinear and, of course, modeled such systems using computers.

The connection of Artificial Life in New Mexico with Los Alamos and with SFI means that Artificial Life is served by many of the sources that fund these two entities—though the field has had virtually no specific support from the lab and has had some difficulty getting funds even at SFI. Artificial Life is sufficiently new that many funding agencies do not know what to do with it. SFI is considered a nonprofit organization and is funded by a number of sources. In 1993, when I began my work, SFI's annual budget hovered around $3.4 million (by the time SFI entered 1997, it had increased to about $5.5 million). A large portion of money comes from the private-sector MacArthur Foundation and from the U.S. public-sector National Science Foundation, Department of Energy, and Office of Naval Research. The Institute also maintains what it calls a business network, and contributors include Citibank/Citicorp, Coca-Cola, Hewlett-Packard, Intel, Interval, John Deere, Shell International B.V., Xerox, and a variety of financial advising companies. Member companies are expected to give at least $25,000. This entitles them to receive SFI publications, to send one company member to SFI workshops and conferences, and to maintain an account on SFI's computer network. Through this program, some SFI scientists have taken on consulting jobs with companies interested in using complex systems ideas in manufacture and management. Individuals also contribute to SFI, including many SFI scientists themselves. Early on, the CEO of Citibank/Citicorp became interested in SFI and helped begin SFI's program in understanding the world economy as a complex evolving system. Individual researchers also bring money for specific projects. SFI sponsors a series of lectures for the public in an effort to locate itself in the community and to find new benefactors.

SFI and Artificial Life are sold to funding agencies and to the public as projects in frontier science. This goes some way toward explaining why the history of Artificial Life is often told as an American success story, as the tale of pioneering scientists seeking a new creation. It also helps us to understand why other lineages for the field are ignored in popular and internal accounts. Thinking about biology as continuous with nonorganic process and using computers as models for organisms are enterprises that have been under way internationally for a long time. In part, Artificial Life at SFI is able to move forward without referring to this work because of an ethnocentric amnesia about these precedents (served by a U.S. science education that continually erases the recent institutional and international history of ideas) and because of its distance from big universities. The formation of Santa Fe Artificial Life as a kind of American story of self-made men wedded to a do-it-yourself aesthetic is an important part of this tale.

ARTIFICIAL LIFE SCIENTISTS AT SFI

The core group of people in Artificial Life in Santa Fe—that handful instrumental in leading the first conferences, in writing mission statements, in making sure that Artificial Life had a home in New Mexico—are overwhelmingly physicists from Los Alamos, and almost all floated around their midforties when I did my fieldwork (there are about five or six of these people, all or almost all of them men, depending on who's counting and how). The men who compose the nucleus of this group came of age in the late 1960s or early 1970s and have partaken in various ways of the ethos of the white counterculture of that era. Many experimented with different lifestyles, and each one has a biography that takes him through a landscape of alternative communities, relationships, life paths, and careers. Most protested against the Vietnam War (Langton was a conscientious objector). All tell stories about themselves as questioning authority and make this an ingredient of their narratives about their paths to science. Some have taken circuitous paths to academia, paths made possible by flexibilities in the U.S. university system. Almost all told me stories about processes of self-reevaluation that had led them to see the world as full of alternative possibilities, both for constructing personal identity and for understanding reality itself. I think it would be fair to say that they see convergences between their involvement in "the sixties" and their involvement in a science they take to be about alternative and unorthodox ways of thinking about life. Several came to Los Alamos to work in sections of the lab "outside the fence," in nonclassified, non-military applications research. Chris Langton, the figure most associated with organizing Artificial Life into coherence, exemplifies some of the characteristics of this cohort of scientists. A roughly sculpted man who still wears his hair long and often accents his denim look with turquoise jewelry, Langton came to his graduate study of computer science in his midthirties, after spending much of his youth traveling around the United States doing odd jobs, playing music, hang gliding, and sporadically attending college.

Los Alamos is clearly connected to the technoscientific military projects of the United States, which people usually associate with nuclear weapons. Computers have played a key role in this research and have been used not only for nuclear physics simulations but also for exploring issues in other branches of physics. All the men in Artificial Life who came through Los Alamos have done extensive simulation work. The lab has a long tradition in the simulation of physical systems reaching back to the 1940s, so it is no surprise that these people should have become attracted to this traditional tool. But the connection to computers is not simply institutional; computers have also figured prominently in these people's individual lives. Without exception, they bought personal computers sometime around the late 1970s or early 1980s, just as they were first becoming available.

The anthropologist Bryan Pfaffenberger (1988) has argued that the personal computer (PC) was marketed as a tool for realizing dreams of self-empowerment, self-realization, creativity, grassroots politics, and democracy. Compared to the mainframe computer—a forbidding symbol of hierarchy and cold rationality—PCs were friendly. The PC was in many ways targeted at people who came of age in the 1960s who were looking for ways to square their idealisms and distrust of central authority with the increasingly normal suburban lifestyles in which they found themselves. The PC is an important figure in the story of Artificial Life; many people I interviewed told me that without their own computers, they would never have been able to do their early experiments in simulating self-reproduction and evolution. In a letter written in the early 1980s to a professor in whose work he was interested, Langton wrote, "I purchased a small Apple computer and started studying simulated models of evolving colonies of 'bugs,' which would make copies of themselves by copying tables of information which described their behaviors with respect to a simulated environment" (n.d.). Many also saw a link between their view of themselves as iconoclasts and the ways they appropriated the PC, and some connected their use of the PC directly to countercultural experiences. One researcher of the 1960s generation saw programming artificial worlds as akin to an experience with LSD, "a spiritual experience." Here one could experiment with a variety of different and alternative versions of reality. Larry Yaeger, an Artificial Life researcher in the same age range as the SFI cohort, told me over email, "Only after about a decade of scientific programming on mainframes, minis, and supercomputers did I finally acquire an Apple II+. It was invigorating, exciting and life-altering." The feeling that computers were tailored to the idiosyncratic concerns of budding Artificial Life scientists was fostered by commercial work done to make computers seem user-friendly, even "natural" in operation (the name "Apple" alone summons forth images of the organic world). It is not surprising that PCs should have been written into people's countercultural dreams about new ways of living and of viewing their relationship with the natural world.

But the meanings PCs were marketed with, and out of which they came, did not simply rearticulate dreams of the white counterculture; they also recombined ideas important in mainstream American political culture: individualism, the realization of democracy through electricity and the connection and communication it makes possible, and self-empowerment through mastering new technologies. Some people might cynically say that the proper frame in which to understand PCs is as part of the transmogrification of 1960s counterculture into an ethics of self-help and individualism, of social rebellion into commodified counterculture, and of collective lifestyles into small-company entrepreneurship (see Ross 1991).

The American ideals embedded in the PC are not generic; they also refer us to a culture in which individual autonomy and creativity is premised on a masculine picture of what it means to create and become adult. While the appropriation of PCs empowered women in various ways (in a rather traditional example, allowing some to work at home while caring for children), it was also done in a rather masculine key.

> As Sherry Turkle has shown, many of the men who purchased the first
> home computers . . . bought the computers as a way of acquiring what
> were to them important symbols of prowess and prestige in an increasingly
> secular world. Such self-administered rites of passage are hardly surprising

in a culture where, as Cynthia Cockburn has argued persuasively, technical skill and high-tech artifacts are widely considered to be the exclusive possession (and constitutive symbol) of male maturity, potency, and prestige. (Pfaffenberger 1988:44, references omitted)

The Artificial Life scientists I am writing of might have had reason to be interested in computers beyond desires to enact a practice of masculine accomplishment, but their masculinity is not incidental. Many told me that their wives complained when they stayed up all night programming. One said, "I think she is jealous of the computer." Beneath discourses of counterculturality there can survive quite traditional structures of feeling (see Bourdieu 1972).

Artificial Life scientists self-consciously purvey an image of themselves as rooted in the 1960s. One *Scientific American* article captured this perfectly when the author called an Artificial Life conference, "a Woodstock for computer hackers" (Corcoran 1992:17). The 1960s ethos was remarked on by one of the younger people I interviewed, a computer scientist with training in physics. He said that Artificial Life was begun by a bunch of ex-hippies who hoped to do to science what they felt they had done to society in the 1960s—revolutionize it. He commented that one could see this in how open the field was; people were invited to participate who had no background in biology and who sometimes did not even have a rigorous computer science background. A few were people whose primary experience was designing computer games. An established older biologist, echoing these thoughts, wrote of the third meeting on Artificial Life, "Since access to computers, workstations, and mainframes is fairly widespread, and since the equipment is so universal, this is a rather egalitarian branch of science, as contrasted with those in which very large set up funds and costly equipment are needed. As a result, many of the participants are somewhat idiosyncratic, and others seem downright flaky" (Morowitz 1992:15). Some of these people might also be described as fitting into the first wave of "hacker" culture, a culture of computer aficionados obsessed with building and understanding unruly systems, often to a degree that sacrificed their sleep and social life. Several people of this profile spoke to me of the addictive qualities of programming and mentioned that in their more euphoric moments they were transported to science fiction fantasies in which they created new worlds in computers. This mixture of science, engineering, art, and fantasy makes Artificial Life an interesting specimen. It is also what has led some to label the field "fringe." Langton has remarked, however, invoking Darwin, that he feels that embracing the fringe is necessary for Artificial Life to get a good pool of efforts from which to select.

In spite of the out-there character of some Artificial Life, or more likely because of it, many younger scientists are attracted to the field and to SFI. Gerald, one of the younger people in the field said to me, "I was originally somewhat taken in by the 'coolness' of ALife research: the hacker mystique, the romance of power associated with high-speed computers, the cachet of robotics, the visual pyrotechnics of ALife videos. But I wouldn't have stayed in the field if there weren't more substance to it." Whereas the core, now middle-aged, group of Artificial Life researchers in Santa Fe came primarily from theoretical physics, the newer, younger people in the field are computer scientists, with little training in physics or biology. Certainly this is not too surprising, since SFI is a research center organized around the computer and since in almost all fields of science there is a growing focus on the use of computational techniques.

SFI undergraduates, graduate students, and postdoctoral fellows take Artificial Life as an already established, if somewhat new, field. This means that they are not as invested in defending it as their immediate elders, and it means that they are comfortable criticizing and rethinking received ideas. The younger researchers do know that associating themselves exclusively with Artificial Life is risky, and most come to SFI to attack problems in more traditional computer modeling. Their résumés recount in deliberately unspectacular language their work with evolutionary algorithms, or with less controversial but still novel search techniques. Some consider themselves primarily computer programmers and see Artificial Life as an interesting way to explore and expand their skills in simulation. All of the young people have grown up with computers and take them as a fixture of their everyday world. Like me, many reported that they were excited by Hofstadter's *Gödel, Escher, Bach,* and found compelling the book's ornamented analogies between DNA and computer programs. A few worked with Hofstadter in their early graduate years. Virtually none had any training in biology; most relied on memories of high school and popular biology texts or on conversations with biologists to implement their Artificial Life systems. Some might also be classified as matching the hacker stereotype; a few have histories of playing Dungeons and Dragons (a game in which one creates a medieval fantasy world for one's friends to inhabit and explore) and of designing their own video games. Younger scientists at SFI often enjoy mathematical games and music and like to talk about whether physics in science fiction books is realistic.

While many young people had a profile not far off from what one might expect for those doing cutting-edge computer work, there were a few who came armed with a highly ironic attitude about Artificial Life. These people had read a good deal of cyberpunk literature, a genre of science fiction whose subject is dystopic futures in which multinational biotech and computer corporations have reshuffled the ways humans and machines are connected and constituted. Younger folk immersed in this literature enjoyed a certain reflexivity about the ambiguous resonances of SFI work, its troubling connections—institutionally and financially—with Los Alamos, and the ways simulation was a perfect postmodern tool for relativizing, reimagining, or reinforcing the realities we inhabit. Whereas the fortyish people were aware of the ethical dilemmas that might be produced by Artificial Life, these younger people had a much more profound sense of how polyvalent ideas might be, how even the most ecologically correct idea might be appropriated by multinational capitalism. Whereas older people invoked morality tales like *Frankenstein* and urged caution in the creation of artificial life-forms, many younger people focused on how Artificial Life technologies might be used for the nefarious purposes of military simulation, for waging wars by remote control, or for organizing, searching, and maintaining IRS and credit card company databases. At the same time, they saw the potential for novel and subversive uses of these technologies, envisioning the possibility of new modes of embodiment, new transgendered and transorganismic subjectivities and cyborg sexual practices. A few read magazines like *Mondo 2000* and *boing boing* that track the multiple and ironic mixings of technoculture, art, science, psychedelia, and countercultural criticism.[14]

For the most part, the younger group at SFI were fairly liberal in their politics. Some reported that Artificial Life should help us to think more ecologically, that computer simulations can help us to recognize the essential interdependence of all life. Here Artificial Life is inspirited with the environmentalist zeitgeist that began to inhabit popular politics in the 1980s. Gerald said that his politics and his interest in Artificial Life were entwined, and here I want to flag how some people make explicit the diverse cultural threads that lead them to Artificial Life—even if, in the end, they argue that it is science that grounds their beliefs. I include an extended quote from an email interview with Gerald because I found him an articulate spokesperson for a set of positions I heard gathered together in many other younger people's words.

> My libertarianism, environmentalism, respect for ordinary human behavior, respect for free markets, feminism, respect for animal rights, fascination with evolution, and fascination with simulation all grew up more or less simultaneously. They seem to be natural complementary ideas to me. My main relevant social concerns are (1) fostering an appreciation of biological complexity so humans don't permanently screw up the environment and extinguish too many species, and (2) fostering a more detached, self-aware, ironic attitude towards our own evolved mental adaptations, emotions, motivations, concepts, etc., so we don't take them so seriously. I focus on investigating our evolutionary heritage because I think it provides the *only* post-religious, post-metaphysical foundation for a concept like "universal human rights." There is a kind of wonderful tension between the notion of human artificiality and the notion of biological nature. It's a productive tension because I think it creates a conceptual or imaginative space where the complexity of biological systems can be compared directly to the complexity of human-engineered systems. And the human systems almost always end up looking *ludicrously* simple, inadequate, inelegant, and clumsy by comparison. By allowing such comparisons, my hope is that the sort of rationalistic, technophilic scientists that get drawn into computer simulation will develop a deeper respect for animals and ecological systems. So I see artificial life as potentially a way of converting "hard scientists" to a greater ecological awareness and sensitivity.

Ken Karakotsios, also a fairly young scientist, told me in one interview, "I believed, and still do believe, that if people can play with ecologies, they'll get a better sense of how their own behavior can become magnified and affect the planet's ecosystems. I also believe that by playing with evolution, people can develop a larger sense of what nature is all about."

Most of those in their fifties and sixties who have become associated with Artificial Life come from rather traditional computer science and have an interest in Artificial Life because they see it charting new approaches to Artificial Intelligence. They have been persuaded that expert systems that simply represent knowledge have limited use and are ultimately too rigid. Artificial Life has been sold as a successor to Artificial Intelligence, as a computational practice that will begin to build intelligent agents from the bottom up. There are a few who remain skeptical of the role of simulation in science; they are not happy that new mathematical tools may need to be developed to understand simulated worlds. They still have a commitment to a "real" world in which things can be tested, and they are very uncomfortable with the notion that simulations can provide a middle ground between theory and experiment, that simulations might be considered alternative worlds (see Casti 1997). Many are more interested in what Artificial Life will do for computation than in what sorts of questions it will answer for biology, which they consider to exist in the world *outside* computers.

There are other ways to tell the story of who comes to the overlapping worlds of Artificial Life and SFI, and a social history would look different if started elsewhere. Many Artificial Life scientists from Europe come through the Institute and have diverse opinions and effects on research at SFI. A few are interested in holistic approaches in the discipline and hope that Artificial Life might show us how organisms and ecologies are interdependent. Some think that Artificial Life is perfectly poised to go up against neo-Darwinian orthodoxy, emphasizing chaos and indeterminacy against the teleological and ideological stories of progress embedded in most evolutionary simulations. At the same time, they see old neo-Darwinian formulations being reinvented in Artificial Life and have definite opinions about why this is happening and how it resonates with barely secularized Protestant U.S. political culture. My final chapter contains closer writing about European arenas of Artificial Life research. I close now with a description of events that took place at a major Artificial Life conference in Santa Fe, picking out moments that speak to the social constitution of the Artificial Life community as it exists in the United States and as it often coalesces around SFI.

ARTIFICIAL LIFE III

In the summer of 1992, an enthusiastic assembly of some five hundred computer scientists, biologists, chemists, physicists, and others converged on the Sweeney convention center in downtown Santa Fe for the third U.S.-sited conference on Artificial Life. For a community dispersed around the United States and around the globe, this is an event that can engender feelings of solidarity. It is a ritual that creates a sense of cohesion; it is a lens through which people can see what they have in common and how they differ.

When I arrived, the convention center was abuzz with energy, as people milled about, waiting for talks to begin. The crowd was young, and some people had a decisively tattered appearance. The gathering had the feel of a science fiction convention, a rock concert, and a scientific conference all swaddled together. People had a sense that they were participating in something cool, as was evidenced by long lines in front of the booth that sold black Artificial Life III T-shirts. The T-shirts displayed a beetle transforming into an insectoid robot (see fig. 2). I purchased the T-shirt myself, and was surprised at the uneasy reactions of my nonscientist friends when they saw it on my body. Many pointed out that it was the sort of shirt that only little boys into insects and robots could love. This reminded me then, and does now, that the conference was overwhelmingly populated by men.

Nestled between and alongside the many talks given at the conference was a seminar on how to build a simple robot. Managed by people from MIT, its aim was to give people the tools they needed to construct robots using Lego, the logo programming language, and circuit boards. Tables were arrayed with electronic detritus, and the small conference room looked like a sixth-grade science classroom retrofitted to speak to the desires of Radio Shack frankensteins. The seminar depended on participants having some familiarity with Lego and home electronics. Before people were released to plunder the Lego and circuit boards to make their own R2D2s, however, a video was shown to illustrate what sorts of robots could be built. It showed MIT-built robots in competition to perform various tasks. In one sequence, robots vied with each other over a collection of Ping-Pong balls while their human makers cheered them on. The event had the character of a cyber-cockfight, with programmers' prestige deeply bound up with the play and performance of their creations. This segment of video was accompanied by the Billy Joel song "Pressure," lending a distinctly masculine sound track to the robot competition it recorded. The video told us that the competition had a traditional "heats and elimination" structure, just like a sporting event.

As the robot builders toiled, people in the main auditorium gave and heard an avalanche of scientific talks on Artificial Life. Several focused on robots, some on bioengineering, but most centered on computer simulation of evolutionary dynamics. Larry Yaeger spoke on his artificial world, PolyWorld, and of how he had evolved populations of polygon-shaped creatures that ate, mated, jogged, and sometimes gobbled each other up (see fig. 3). Yaeger, with his long gray hair and T-shirt, looked like a beardless hacker rendition of the fatherly God of Judeo-Christianity. Younger Ken Karakotsios spoke of his creation, SimLife, a software toy for exploring evolution. He called SimLife "ALife for the masses" and argued that it could help people understand the inter-

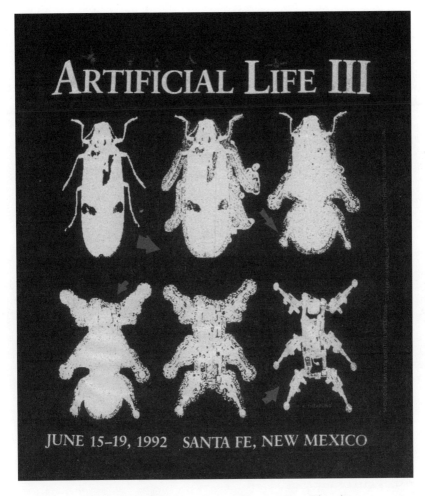

Figure 2.   Artificial Life III T-shirt. Copyright 1992 Kurt Thearling.
Reproduced with permission.

connectedness of things in ecological systems. "If people understand
complex systems," he suggested, "they might think twice about buying
a gas-guzzling car." Charles Taylor of UCLA spoke of using computer
simulations to model and control the growth of malaria-carrying mos-
quito populations in California and Mali. Craig Reynolds spoke of how
flocking and schooling behaviors had emerged in a program populated
with digital models of birds called "boids" (see fig. 4), and he dazzled
the audience with images of flocking bats he had helped generate for the

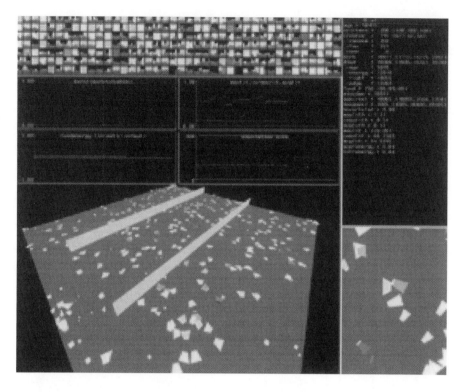

Figure 3.   Screenshot of PolyWorld, Copyright 1992 Larry Yaegar.
Reproduced with permission.

popular film *Batman Returns* (more recently, he has worked on the
wildebeest stampede in Disney's *Lion King*). And Tom Ray spoke on new
results in Tierra.

   One of the most popular talks, "A Case for Distributed Lamarckian
Evolution," was given by the computer scientist David Ackley (who
coauthored the associated paper with Michael Littman). He declared
that Darwinian evolution, which disallows evolution within an organism's lifetime, sounded "stupid" from a computer science perspective.
Computer life-forms should have no need to wait until death to evalu-
ate whether their genetically encoded survival strategies are fit; they
should, given direct access to their genetic code, modify their genetics as
they go along. Unlike Darwinian evolution, Lamarckian evolution
(named after the pre-Darwinian naturalist Jean Baptiste de Lamarck)
could allow organisms to transmit characteristics acquired during their
lifetimes to their offspring. Noting that in any evolutionary process you

Figure 4.   Flock of "boids" negotiating a field of columns. As described in Craig W. Reynolds, "Flocks, Herds, and Schools: A Distributed Behavioral Model" (Proceedings of SIGGRAPH 1987), *Computer Graphics* 21(4):25–34. Reproduced with permission.

want to "squeeze out the idiots," Ackley described evolution as a process of optimization and argued that this optimization could be more efficiently done when organisms "had no bodies." Ackley cut an amazing figure. Speaking bullet fast with a brilliant sense of comic timing, he lurched around in his bulky black-clothed frame, his long black hair and pointy beard moving to punctuate his points. Ackley talked about programs that "had sex," that could be described as related like "father" and "son." And he spoke of how the diversity maintained in marginal populations could be important for evolutionary innovation. As he put it, in a sentence that drew cheers from an audience that liked to imagine itself as somewhat crazy, "Instead of viewing the lunatic fringe as a sad necessity, we see it as a repository of innovation." (Just what the difference was between lunatics and the idiots Ackley had mentioned earlier was not made clear.)

The roboticist Mark Tilden, giving a talk that seemed engined by a desire to escape his body, tickled the audience with robots created from old calculators and radios. His presentation included a discussion of the "disadvantages of biology," among them "the 'emotion' thing." Tilden wore a tight ponytail and resembled nothing so much as a thirty-year-old kid, all grown up and continuing childhood fascinations with robots. He showed a set of photographs of his new creations, and he toted them around as might a proud parent. He beamed as he showed one of them, "This creature right here is just a few weeks old." Tilden's visions of what robots might do sounded as though they were generated to speak to an audience of ecologically minded survivalist bachelors. He said, "We want robots that are autonomous and that can survive if cut off from their buddies. Robots to clean up oil spills, terraform Mars, and clean the bathroom and vacuum." These robots would be more robust than those constructed following the science fiction writer Isaac Asimov's laws of robotics. Asimov's laws declare that robots' first duty should be to serve and protect humans. "Tilden's Laws of Robotics" would be "1. Protect thine ass, 2. Feed thine ass, and 3. Look for better real estate."

While there were moments when Artificial Life practitioners fairly bludgeoned the audience with images of themselves as out-of-control boys playing at single-handedly fathering new life, this interpretation misses many other meanings in play at this gathering. Some saw Artificial Life as fostering a more participatory view of life and ecology. Several spotlighted their concern with ecology and with Artificial Life as a path toward thinking of nature in a less dominating mode, as a tool for recognizing that we are always part of nature, and most so when we presume to design and interpret it. These people, most of them men, sometimes advocated a sort of Gaian, eco-feminist vision of the world, one that went to another extreme, assuming not a hostile nature but a nature acting with good intentions.

One highly visible person was a white environmental activist and computer scientist in his thirties, Howard Perlmutter, who had just returned from the 1992 environmental summit in Rio de Janeiro. Though not a

scheduled speaker, Perlmutter rose from the audience repeatedly to pro-
mote ecological points of view and to argue with scientists he felt were
ignoring potentially ecological morals emergent from their own research.
Over a special lunch he organized, and which I attended, he read biodiversity treaty statements and talked about differences between North
and South interpretations of environmental crisis. He read the Malaysian
critique of the Rio conference, which pointed to the place of colonial relations, northern overconsumption, and unequal political-economic and
industrial relations in exacerbating environmental problems. He tied environmental politics to ideas about "emergence" in Artificial Life and
took Artificial Life's focus on the generation of complex behaviors from
local rules to be aligned with a politics of acting locally and thinking
globally. He chided Artificial Life researchers for not recognizing what
was obvious to him, that a definition of life would be more of a value
judgment than an empirical question. When he intervened, he was often
dismissed, perhaps in part because his self-presentation bordered on the
New Age; he wore a tie-dyed shirt and casual, colorfully striped
Guatemalan pants. He did not highlight the fact that years earlier he had
started a software development company in Santa Cruz called Soft-
weaver, which was devoted to a practice of programming wedded to a
holistic philosophy of life and electronic communalism (see Stone 1995).
Many people at the conference simply regarded him as a "politically correct" ideologue.

So it was that the Santa Fe conference attracted a group of people with
a certain eclecticism, but a group that connected well with the SFI Artificial Life group I have described, a group committed to computer
worlds, but also on occasion to ecological agendas and concerns. At a
few moments during the conference, I asked researchers to characterize
the differences they saw between the researchers present. Only a few
touched on the sociological aspects I have sketched here; most focused
on more internalist distinctions. One Institute-affiliated man said that as
far as he was concerned, most differences between people boiled down
to differences in the questions they were asking: Artificial Life meant
something different to a roboticist, a computer scientist, and an evolutionary biologist. Being interested in metabolism, self-reproduction, or
flocking will lead one to different issues and to different conclusions
about what is essential to life. When this person went to his first Artificial Life conference, he said that he thought the robotics people were not
doing Artificial Life at all. Robots could not self-reproduce, he said, and

so should not count as alive. Gradually, it occurred to him that people were just focusing on different aspects of living systems. He saw Artificial Life III as sensibly organized to bring people with these different approaches into dialogue. But like many at Santa Fe, he thought that simulation was really the heart of the enterprise.

The five-day conference wound down, participants exhausted. Chris Langton sent us on our way, saying, "So these are the survivors. You all go on to the next generation to have offspring." This was a joke, of course, but suggested that those who attended the conference would reproduce the ideas presented there—reproductions that would take place, among other sites, in the virtual worlds of computers. It is to the topologies of these silicon second natures that I dedicate my next chapter.

TWO

ACHILLES: It's most perplexing how the characters in my dreams have wills of their own, and act out parts which are independent of MY will. It's as if my mind, when I'm dreaming, merely forms a stage on which certain other organisms act out their lives. And then when I'm awake, they go away. I wonder where it is they go to . . . AUTHOR: They go to the same place as the hiccups go, when you get rid of them. . . . Both the hiccups and the dreamed beings are software suborganisms which exist thanks to the biology of the outer host organism. The host organism serves as a stage to them— or even as their universe. They play out their lives for a time—but when the host organism makes a change of state—for example, wakes up—then the suborganisms lose their coherency, and cease existing as separate identifiable units.

DOUGLAS HOFSTADTER, *Gödel, Escher, Bach*

The Word for World Is Computer

ARTIFICIAL LIFE SCIENTISTS consider computers to be worlds in a variety of senses. Often they mean that computer simulations can be seen as self-contained, self-consistent symbolic models that generate objects and behaviors corresponding to things in the real world. Increasingly, they mean that computers are literally alternative worlds or, more ambitiously, alternative universes. Making sense of this contention is important, for it makes possible the claim that real life-forms can exist in the computer. I want to look at this contention stereoscopically. On one side, I want to inspect the initiate's view that computers can be life-sustaining worlds. On the other, I want to examine the cultural resources "outside" science that researchers use to construct computers as alternative realities. I am interested to show how elements from complex systems science, computer science theory, and quantum mechanics fuse with elements from Judeo-Christian cosmology, science fiction, an American frontier imagination, a psychedelic imaginary, and televisual common sense to persuade many Artificial Life scientists that computers can be regarded as their own worlds.

The title of this chapter is a play on Ursula K. Le Guin's *The Word for World Is Forest* (1976), a tale about a tribe of forest-dwelling people who use their dreams as a resource for guidance in the waking world. The Le Guin story is a fanciful one that imagines a people who exist outside of the stream and stress of history, so it may be careless to tweak its title to speak about turn-of-the-millennium scientists who see themselves wizarding up new worlds, new universes, and new creations in which to carry out the future missions of evolution. There is, however, a certain romantic incandescence to the project of Artificial Life, to practitioners' claims of being technologically in tune with nature, that leads me to discern similarities between Artificial Life researchers and Le Guin's mythical Athsheans. Artificial Life workers, like Athsheans, draw on a set of culturally specific dreams—of transcendence, discovery, creation, colonization—to manufacture the worlds they interpret.

For many people I interviewed, a "world" or "universe" can be understood as a self-consistent, complete, and closed system governed by low-level laws that support higher-level phenomena, which, while dependent on these elementary laws, cannot be simply derived from them (here "low" refers to physics and "high" to chemical, biological, and social phenomena).[1] Put into the language of the sciences of complexity, a world or universe is a dynamical system capable of generating surprising emergent properties. If we accept this definition, then computational systems reasonably count as worlds or universes. This is the view taken by the Artificial Life scientist David Hiebeler in an article entitled "Implications of Creation," in which he writes, "Computers provide the novel idea of simple, self-contained 'artificial universes,' where we can create systems containing a large number of simple, interacting components. Each system has its own dynamics or 'laws of physics'; after specifying those laws, we set the system into motion and observe its behavior" (1993:2). (See fig. 5, a sketch of a simulation that Hiebeler and others were working on during my time at SFI.) Larry Yaeger drew the similarity between worlds and universes and computer simulations for me rather tightly: "Worlds and universes are complex processes, based on fixed, low-level principles. Computer simulations are complex processes, based on fixed, low-level principles." Tom Ray has said of Tierra, "Tierra is Spanish for Earth. I called it that because I thought of it as a different world that I could inoculate with life. It's not the same world we live in. It's not the world of the chemistry of carbon and hydrogen. It's a different universe. It's the chemistry of bits and bytes" (Ray in Santa Fe Institute 1993).

I did meet a few scientists who were skeptical that computers could be worlds or universes, and I record their doubts before I proceed, because they are scientific permutations of the idea that computers are devices, not places. One older physicist I spoke with grimaced when I asked what he thought of some researchers' predilection for calling computer simulations "artificial universes." He said that the universe was the set of all matter that exists, and he asked, as if greatly annoyed at people's

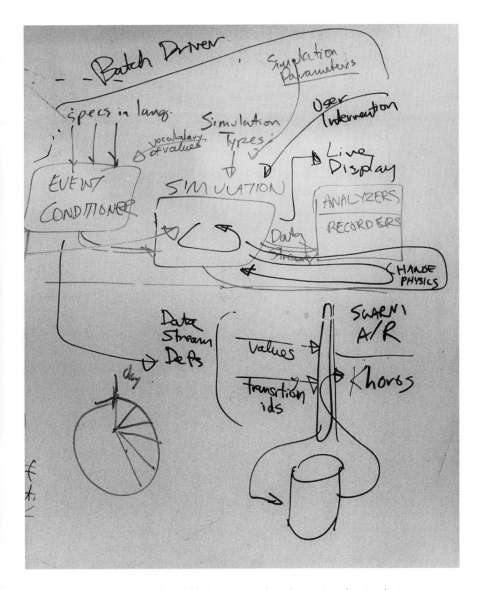

Figure 5.   Whiteboard scribble laying out the schematics of a simulation.
Note the "change physics." Photo by Stefan Helmreich and Chel Beeson.

stupidity, how a subset of that, namely a computer, could be a universe.
One older biologist was impatient with the idea that computers might
be self-contained universes governed by alternative physics and chemistry. He said he was not interested in understanding life-forms that

violated the first and second laws of thermodynamics, life-forms that did not exist in the material universe he knew. He said, "If I can't take it to the periodic table, I'm not interested."

Many nonscientists I spoke with at SFI and elsewhere expressed what I think is a common view, that computers are tools or telecommunications devices. Chris Langton shared with me the email comments of someone skeptical that computers were worlds. "What is a computer, anyway?" this person asked. "Basically it is a machine—a tool for performing computation. Given that it is a tool for computation, it computes. That's what it does and that's all it does. Granted, it's wondrously general and can be used to do many things, but bits is bits. So, given that the computer is a computation machine, it can be used to model and simulate. But to say that it can support life would be claiming that it is in some way equivalent to the universe that supports all life as we know it." Many Artificial Life people do indeed claim that computers can support life and can do so because they *are* in some way akin to the universe we know. How has this common sense come into being?

COMPUTERS AS "MICROWORLDS"

Computers have been used for quite a while to simulate aspects of the world we inhabit. Los Alamos scientists used them early on to simulate nuclear reactions, and the U.S. military has also used them to model nuclear war scenarios. Simplified computer models of the world became a key component in Artificial Intelligence research in the early 1970s, as scientists attempted to provide programs with representations of fragments of the world about which the programs were to make reasoned decisions (see Winograd 1972; Dreyfus 1979; Edwards 1996). These representations came popularly to be called "microworlds." The historian Paul Edwards (1996) has argued that computer microworlds are also "closed worlds," spaces in which all variables that affect a situation are well-specified while other variables are considered irrelevant.

The manufacture of the computer as a closed world was linked to the growth of computer models in the military, which favored computer simulations of conflict that simplified international relations into formalizable and well-contained game scenarios. If the political world could be modeled as a closed, consistent system, so the reasoning went, perhaps these models could be used as technical tools to control it. Edwards has argued that the military origins of computer modeling set the stage for understanding computers as surrogate worlds, and he has written that the image of computer as "world" has served as a powerful attraction for many people:

> What gives the computer [its] "holding power," and what makes it unique among formal systems, are the simulated worlds within the machine: what AI [Artificial Intelligence] programmers of the 1970s began to call "microworlds," naming computer simulations of partial, internally consistent but externally incomplete domains. Every microworld has a unique ontological and epistemological structure, simpler than those of the world it represents. Computer programs are thus intellectually useful and emotionally appealing for the same reason: they create worlds without irrelevant or unwanted complexity. In the microworld, the power of the programmer is absolute. Computerized microworlds have a special attraction in their depth, complexity, and implacable demands for precision. The programmer is omnipotent but not necessarily omniscient, since highly complex programs can generate totally unanticipated results. . . . This makes the microworld exceptionally interesting as an imaginative domain, a make-believe world with powers of its own. (1996:171–172)

Edwards captures the textures of computing that have attracted many people to Artificial Life. He maintains that the microworlds of computers have been particularly appealing to men:

> For men, to whom power is an icon of identity and an index of success, a microworld can become a challenging arena for an adult quest for power and control. Human relationships can be vague, shifting, irrational, emotional, and difficult to control. With a "hard" formalized system of known rules, operating within the separate reality of a microworld, one can have complexity and security at once: the score can always be calculated; sudden changes of emotional origin do not occur. Things make sense in a way human intersubjectivity cannot. (1996:172)

Helena, a younger scientist I interviewed, agreed with this view and said that science promotes habits that are gendered masculine:

> I think scientists are usually just sort of socially disabled, so they enjoy forming systems in which they don't have to pay attention to the real niceties of complexity, real social interactions, and things that they're not good at. There's also something in the training of physicists and computer scientists that makes them assume that everything will reduce to a certain level of simplicity. So, they assume without looking at real systems—real biological or ecological systems—that everything will reduce to this particular level, so they figure they might as well go and make something like that, because then they'll have control over it. Often they're these men who think, "I'll rule my own kingdom then."

One man reflected self-consciously that he built simulations because he wanted to get away from a world that felt difficult to control. He said to me, "I didn't like chemistry because it involved measuring the real world, and I've never been a big fan of the real world. I'd much rather make my own world on the computer and then measure that." Statements like this are enabled by a normative kind of Euro-American masculinity that consists of emotional disengagement, escape and autonomy from others, calculative rationality, and a penchant for objectifying, instrumentalizing, and dominating the world. This is a broad-brush characterization but useful for understanding how computers have been forged into worlds in the crucibles of the military, Artificial Intelligence, and masculinity. As I have noted, many men in Artificial Life purchased PCs as part of their personal rites of passage, rites especially connected to mastery of technical-rational skills since these men were all scientists.

Of course, to boil down to masculinity people's perceptions of computers as worlds is far too simple. Masculinity is by no means monolithic. It is crosscut by complex histories of race, sexuality, class, and nationality. Gender names an analytic category that describes sets of characteristics and dispositions, often popularly believed to connect to "natural" sex difference, which is used to distinguish between "men" and "women." Such characteristics are nonetheless never the exclusive property of one gender or the other. One woman I spoke with expressed what some might see as a typically "masculine" motivation for using computers: "Having the world at your control is something that has a lot of appeal. You can figure the whole thing, with the parameters you want. It gives you this sense that everything's fine and under control."

In many of the artificial worlds of Artificial Life, however, everything is not under control. Researchers actively hope that their worlds will surprise them with lifelike dynamics. In the psychologist Sherry Turkle's (1995) terms, an aesthetic of "hard mastery," in which programmers seek to completely comprehend and structure programs, is giving way to one of "soft mastery," in which people encounter and appreciate computers as dynamically changing systems only partially available to complete understanding. The masculinist imperatives of standard Artificial Intelligence—rationality, objectivity, disinterestedness, and control—yield to what are commonly characterized as more "ecological" programming techniques. This shift follows changes in technologies, from the modernist top-down logics of IBMs to the bottom-up, user-friendly opacities of computers like Macintoshes. Turkle argues that new modes of computing are more appealing to women, who have traditionally been trained to value negotiation, relationship, and attachment (see Gilligan 1982). But this is only part of the story. Many men in Artificial Life see themselves as refiguring their masculinity as they work with modes of computation that mimic "nature." In a conversation with two younger heterosexual men about how their gendered subjectivity might be implicated in their science, they told me that Artificial Life allowed them to express and work with a side of themselves that was more intuitive, perhaps more stereotypically "feminine." This way of putting things shores up the category of gender even as it purports to erode it, keeping "femininity" as a stable resource that men might mine to "broaden" their intellectual work. As Andrew Ross writes, "Patriarchy is constantly reforming masculinity, minute by minute, day by day. Indeed, the reason why patriarchy remains so powerful is due less to its entrenched traditions than to its versatile capacity to shape-change and morph the contours of masculinity to fit with shifts in the social climate" (1995:172). The philosopher Judith Genova notes in "Women and the Mismeasure of Thought," "Perhaps it is no accident that men become intuitive and holistic just at the time when computers can successfully simulate logical, analytic thought" (1989:212). As masculinity is reimagined in the practice of Artificial Life, we might do well to ask, with the semiotician Teresa de Lauretis: "If the deconstruction of gender inevitably effects its (re)construction, the question is, in which terms and in whose interest is the de-re-construction being effected?" (1987:24). The ways gender is getting recoded in Artificial Life scientists' subjectivities troubles any simple attachment of Artificial Life aesthetics to masculinity, old or new, but it is nonetheless important to keep masculinity in view, especially when male researchers speak of themselves, as I discuss later, as Godlike figures "inoculating" computational matrices with self-replicating "seed" programs.

The notion "microworld" has largely been superseded by "artificial world," signaling a shift from understanding computer worlds as "toy" worlds to seeing them as realities in their own right. Certainly, terms such as "Artificial Intelligence" and "Artificial Life" invite these neologisms, but there is more to it than this. How is it that people have come to believe that computer worlds are not just good representations of worlds but might be real worlds in themselves? Edwards uses the word *world* metaphorically and assumes the people he writes about do too. But some Artificial Life scientists are coming to take the word literally.

SCIENTIFIC WARRANTS
FOR CONSTRUING COMPUTERS AS WORLDS

COMPUTERS AS NONLINEAR AND COMPLEX SYSTEMS

I want to back up a bit more into history to revisit high-speed computing and simulation as they developed at Los Alamos during and after
World War II. Los Alamos is important for my story because much SFI
common sense about computers comes directly from the lab. The historian of science Peter Galison discusses how computer simulations were
employed by nuclear weapons researchers at Los Alamos and elsewhere
for problems too complex to solve analytically and impossible to investigate experimentally. Computer simulations of microphysical events like
nuclear reactions came to occupy a liminal place between theory and experiment. On the one hand, they resembled theory because they set into
motion processes of symbolic manipulation. On the other hand, they resembled experiments because they exhibited stable results, replicability,
and amenability to error analysis procedures. "Data" generated by simulations could be given the same epistemic status as data from "real" experiments (Galison 1996:142–143). On a deeper level, the simulations
in question—those using pseudorandom numbers as starting points for
the emulation of physical processes—were seen to share a "fundamental affinity" with "the statistical underpinnings of the world itself"
(1996:144). As Galison notes, "The computer began as a 'tool'—an object for the manipulation of machines, objects, and equations. But bit by
bit (byte by byte), computer designers deconstructed the notion of a tool
itself as the computer came to stand not for a tool, but for nature itself"
(1996:156–157). Galison claims that Monte Carlo simulations, as these
were called, were assimilated to experiment status in part because the
stochasticity embedded in them was seen as directly analogous to the
stochastic processes that characterized microphysical nature. Theories
could be tested with reference to an artificial reality that was just as good
as the real thing, that was in fact itself a sort of alternative, artificial reality. Simulations could be stand-ins for experiment but more boldly
could be seen as understudies for nature itself.

Researchers in Artificial Life inherit this tradition and epistemology
of simulation. In my interviews, people gave me a more current take on
the ontological similarities between worlds and computers: both worlds
and computers, they said, are dynamical systems, and both can be considered varieties of complex, nonlinear systems. This idea is fed by a new

ingredient that has been mixed into the epistemological stew. Many Artificial Life researchers make the metaphysical claim that computers are like nature but, unlike users of the Monte Carlo, do not point to similarities between the stochasticity of statistical processes implemented on computers and the stochasticity of deep nature. Rather, they see both computers and nature as enacting processes of information transformation. They hold that the universe is a computer implementing transformations of information and that computers, which do this so well, should on this definition be considered universes or worlds. That is what Tom Ray means when he writes, "The computational medium of the digital computer is an informational universe of Boolean logic, not a material one" (1994a:184).

But matter need not be factored out entirely. Steen Rasmussen and others have insisted that matter is self-programmable, that the world we inhabit is really a computation instantiated in the matter that composes our universe (see Rasmussen, Feldberg, and Knudsen 1992).[2] Rasmussen told me in an interview, "We should view matter as being programmable, because that's truly what it is. You can look at the informational and functional aspects of matter and then we can see that it is recombination of matter in all kinds of different ways that creates these functional capabilities that enable living systems to come about." If the universe can be viewed as a computation, researchers say, then computations can be worlds or universes.

CELLULAR AUTOMATA

Many people in Artificial Life have been enamored of a mathematical formalism known as the cellular automaton (CA), which allows a programmer to specify rules for local interaction between "cells" on a latticelike grid and to study the emergent consequences of those rules. Here is a description of the CA from the personal correspondence of Christopher Langton: "Imagine a huge plane, divided up into individual squares by a rectilinear grid, like a checkerboard or a sheet of graph paper. Each square in the grid is a Cell of the array, and each cell can be in one of N states, which can change through time. Each cell's *next* state is determined by its *present* state, and the present state of its immediate neighbors" (n.d). States of cells change according to transition rules that apply to all cells, and using CAs, a variety of simple systems can be simulated. While I was at SFI, people simulated bone marrow cell growth, bacterial growth, the growth of patterns on seashells, fluid dynamics, and the voting patterns of individuals who made decisions based on their neighbors'. One popular use of CAs is in modeling self-reproducing automata. Von Neumann envisioned such implementations, and Chris Langton, in the early eighties, programmed a simple loop that could make copies of itself in a field of undifferentiated cells (Langton 1984). Some people at SFI have been trying to develop CA rules that can perform complex computations (see Mitchell, Crutchfield, and Hraber 1993; Crutchfield and Mitchell 1994).

Because of the generality of the CA formulation, virtually any process that can be algorithmically specified can be modeled using CAs. CAs thus support alternative universes capable of sustaining alternative realities. Langton writes that "the transition function for the automata constitutes a *local physics* for a simple, discrete space/time universe. The universe is updated by applying the local physics to each 'cell' of its structure over and over again" (1989:28). For many researchers, this is not just a "toy" universe. As the Artificial Life researcher Andrew Wuensche writes, "A CA may be regarded as a parallel processing computer. . . . A CA may alternatively be viewed as a logical universe with its own local physics, with [emergent structures] as artificial molecules, from which more complex [emergent structures] with the capacity for self-reproduction and other essential functions of biomolecules might emerge, leading to the possibility of life-like behaviour" (1994:3). Some people I interviewed claimed that our own universe might be thought of as one mammoth cellular automaton. In the journalist Steven Levy's *Artificial Life,* we learn that "some [researchers] would even declare that CAs are sufficiently complex to develop an entire universe as sophisticated as the one in which we live. Indeed, claimed one researcher, we had no proof that this universe in particular was not a CA, running on the computer of some magnificent hacker in heaven" (1992b:58). While I was at SFI, I overheard many jokes and serious comments about how this could be so. One person wrote me in email,

Many years before the term "Artificial Life" was coined by Langton I learned about cellular automata and von Neumann's research and results in "Game of Life" research. From that point on I tended to think of the physics of our universe as a 3-D cellular automaton (even though I know such models tend not to have gravity, other fields, distortions of space, quantum effects, etc.). I view myself as a pattern in a CA world—one in which motion is just an illusion.

Many of the core people in Santa Fe Artificial Life began experiments at home with CAs, and some were fascinated by the most popular of CA formulations, John Conway's "Game of Life," in which the switching off and on of cells results in patterns that can look like shapes growing or traveling across the screen. Ken Karakotsios told me in an email interview, "By chance, my work in computers got me exposed, in a peripheral way, to Cellular Automata. What I like about CA is the very thing that makes them so unusual in the clockwork world of computers: they harbor the unexpected; emergent behavior that's not designed into them." CAs are very compelling metaphors for thinking about the generation of complex emergent phenomena, in part because they are visually surprising, and the consequences of low-level rules really are quite unpredictable.

THE PHYSICAL CHURCH-TURING THESIS

Because SFI is so focused on computation, many SFI scientists subscribe to versions of the physical Church-Turing thesis: the notion that any physical process can be thought of as a computation and that therefore any physical process can be re-created in a computational medium. In the 1930s, the mathematician Alan Turing proposed that any human cognitive process that could be described algorithmically could be translated into a sequence of zeros and ones and could therefore be implemented on a computer. After it was shown that the lambda calculus (a recursive mathematical system that allows functions to act as objects of other functions) of the logician Alonzo Church was equivalent in power to the Turing machine (itself equivalent to CAs), the Church-Turing thesis was formulated, stating that all reasonable computational processes are equivalent to discrete, digital models of computation. The physical Church-Turing thesis figures the universe as such a computational process, as a physical system that converts inputs, or initial conditions, into outputs, the system's final state.

QUANTUM MECHANICS

The notion that computational processes can lead to unpredictable complexity resonates with modern physics' view of the universe as a stochastic system. One scientist suggested to me that researchers refer to their simulations as worlds because many of them participate in the culture of twentieth-century theoretical physics, including quantum mechanics, which sees reality as synonymous with mathematics (an idea certainly not without precedents in classical physics). According to twentieth-century physics, the world can be understood as a quantum mechanical system describable by suitably detailed equations. But if one can describe an object or a phenomenon in terms of equations and can do so completely enough that it becomes impossible to distinguish an object from its description, then it makes sense to say that the world really is mathematics.[3] As this scientist put it, "If you cannot ask a question of an object that distinguishes it from its description, then the immediate question is, is there any difference? to which the answer would likely have to be no."

Many (certainly not all) people trained in physics believe that there is no observable property of any system that in any meaningful sense has a reality extending beyond the mathematical equations that can describe it. This belief allows people to say that their computer, since it is a medium for the expression of mathematical phenomena and can solve quantum mechanical equations, is an alternative world. Some scientists, eager to make their computers more like quantum mechanical reality, hope to fashion cellular automata that are less like a caricature of classical physics and more like quantum mechanics. Others envision quantum computers, machines built on and exploiting quantum effects like superposition.[4]

According to quantum mechanics, the only thing that truly exists in the universe is pattern; the substance that supports these patterns is, in some fundamental sense, inessential (in a familiar example, our bodies' molecules, etc., are being replaced constantly, so that we are not so much our matter as the pattern of our matter). This idea that universes are just patterns allows people to claim that "universes" can be transported from one kind of computer hardware to another; universes are, as the computer science jargon goes, "portable." Because many of the founders of Artificial Life are theoretical physicists, this view of reality as physical, quantum mechanical pattern suffuses their thinking and influences how they construe their computer simulations.

This view is common enough that it informs the kinds of jokes people make. One day I was strolling down a Santa Fe street with two younger researchers who were well versed in theories of computation and quantum mechanics. As we walked, one remarked that the sky looked pretty, and he said something like, gee, great software, look at the resolution of those clouds. This joke was about the real world as quantum mechanical computational simulation and about how our universe might just be the product of a cosmic corporation dedicated to selling us images of nature modeled after our fantasies of what nature should be (or after our obsessions with what the New Mexico desert sky should be). This sort of joke was characteristic of some of the younger set of SFI researchers. In telling these jokes, these people ironically and self-consciously commented on their shared cultural knowledge of arcane scientific understandings of the world at the same time that they made it clear that they were either agnostic about what the world really was or willing to accept the possibility that the whole thing might just be a big CA.

Ideas about computers as stochastic or complex nonlinear systems, formulas drawn from cellular automata theory, the physical Church-Turing thesis, notions from quantum mechanics—all of these are scientific resources that Artificial Life researchers can use to think of computers as ontologically like nature and hence capable of being worlds. That theories of computation and theories of physics are articulated together is significant here and locates SFI theoreticians of Artificial Life as trained in or aware of computational spinnings of theoretical physics. The idea that the universe is a computation is so prevalent that it affects the ways some practitioners construct their intuitions about the physical world. Andrew, a computer scientist and Artificial Life researcher in his early thirties said to me, "The universe is a computation of which I am a part. It's the information patterns of things that make them important rather than anything else. I'm perfectly comfortable . . . with the notion that we're running on some big simulator out there. It seems to me as good as anything."

PHILOSOPHICAL POSITIONS

While there are many scientific warrants for construing computers as worlds, the contention that computers are worlds or universes is still controversial, even in the Artificial Life community. Some philosophers who have jumped into the Artificial Life ring argue that computers are just tools for manipulating symbols and that the fact that they can be systematically interpreted as worlds does not make them worlds. And roboticists argue that the emergent realities of a world cannot be duplicated in a computer program. Mark Tilden said in a talk at SFI, "Simulation is like an eclipse—it's brilliant and often beautiful, but you usually don't realize the damage looking at it has caused until it's too late."

Langton told me of a conversation he had with the philosopher Stevan Harnad about artificial worlds. Harnad said he could imagine an artificial system that produced signs systematically interpretable as a world and that even contained signs systematically interpretable as Stevan Harnad having a conversation with Chris Langton. He said that if such a system were created in a computer, it would not be a world, and entities in it would not be alive. As Harnad reconstructs his position in the journal *Artificial Life,*

> The virtual system could not capture the critical (indeed the essential) difference between real and virtual life, which is that the virtual system is and always will be just a dynamical implementation of an implementation-independent symbol system that is systematically interpretable as if it were alive. Like a highly realistic, indeed oracular book, but a book nonetheless, it consists only of symbols that are systematically construable (by us) as meaning a lot of true and accurate things, but without those meanings actually being *in* the symbol system. (1994:297–298)

Harnad has argued that virtual worlds and virtual life can no more be real worlds and real life than simulated fires can make things burn.[5]

Computationally inclined Artificial Life researchers have a response: things *can* be alive or on fire *with respect to* computer worlds. One researcher told me, " 'Life' can only be defined *with respect to a particular physics.* A computer virus is almost as 'alive' as a real virus (not yet, but close) but only within the physics of the computer memory." This notion that computers can contain separate, closed worlds came up again and again, and was one of the rhetorical moves people used to grant simulations their own ontology. One person told me that, with computer simulations, "we have something totally self-contained in the computer." Andrew argued for the distinctness of computational realities by saying that self-reproducing programs were "alive in there" but only "a model out here." He continued, "They're alive with respect to their own universe, their own rules. Wimpy, pitiful little life, but I can't rule it out."

But computational Artificial Life researchers are not content claiming that computers contain only symbolic or informational worlds. Some have been adamant about asserting the physicality or materiality of computers and have used this as a lever in claiming that virtual artificial life is indeed real. The computer scientist Bruce MacLennan has made an argument that many Artificial Life workers have found compelling:

I want to suggest that we think of computers as programmable mass-energy manipulators. The point is that the state of the computer is embodied in the distribution of real matter and energy, and that this matter and energy is redistributed under the control of the program. In effect, the program defines the laws of nature that hold within the computer. Suppose a program defines laws that permit (real!) mass-energy structures to form, stabilize, reproduce, and evolve in the computer. If these structures satisfy the formal conditions of life, then they are real life, not simulated life, since they are composed of real matter and energy. Thus the computer may be a real niche for real artificial life—not carbon-based, but electron-based. (1992:638)

This interpretation has pleased people who are fussy about the distinction between worlds and universes; here computers are figured as worlds that exist in the same universe as the more familiar organic world we know. A computer simulation manipulates structures and patterns of real voltages and so is not purely "symbolic," though its symbolic character is important for how those real events take place (Peter Godfrey-Smith, personal communication).

Part of what is at stake in contending that real artificial worlds and life-forms exist in computers is the claim that computational Artificial Life studies alternative biologies (see Hayles 1996:155–156). Craig Reynolds, most famous for his simulations of bird flocking, defended simulation against the attacks of roboticists at a robotics-focused conference I attended in Brighton, U.K. He claimed that the adaptive programs he was developing differed from robots only in that they acted in a different world. Claiming an ontology for computer worlds was important for Reynolds epistemologically, but also practically; without this claim, Reynolds could not also claim that the agents in his program were adaptive. And without this claim, computational Artificial Life could not pretend to offer alternative biologies to expectant scientists in theoretical biology.

But the fact of materiality does not solve the puzzle of the role of symbols. When MacLennan writes, "If these structures satisfy the formal conditions of life, then they are real life, not simulated life, since they are composed of real matter and energy," he not only retreats from the materiality he is trying to assert (by stating that what matters is form) but also ignores the fact that whatever the "formal conditions of life" are, they will be defined in language. The philosopher Brian Cantwell Smith (1994) has argued that computers' very physical existence is completely enmeshed in our social and linguistic world, that how disk drives, windows programs, file systems, and random-access memory (RAM) caches work is the result of many scientific, cultural, and economic decisions. To treat computers as entities that come to us straight from nature ignores the work and decisions that have produced them, that allow us to contemplate them as worlds. Material practices—of which language is one—have made and will make or unmake computers as worlds or as niches for artificial life. Chris Langton's argument that computer programs that cannot be linguistically distinguished from known life must be considered alive—the position he took against Harnad—recognizes that language is an essential technology for materializing life. I would side with Langton against Harnad but say that Langton does not adequately recognize the historical specificity of his own use of a language that names "life" as a process that dwells in both organic and electronic entities. As the rhetorician Richard Doyle points out, there is nothing about computers in themselves (their increased speed, their capacity to set math in motion, their informatic logic) that forces us to see them as worlds or as homes for artificial life (1997b:110). Rather, he argues, our ways of seeing and working with them are structured by a kind of "rhetorical software" that allows us to enliven them with narratives fished from the reservoirs of our culture.

CULTURAL RESOURCES
FOR CONSTRUCTING COMPUTERS AS WORLDS

NATURE AS LAW-GOVERNED

Many Artificial Life researchers use "world" or "universe" as synonymous with "nature." The Western scientific construction of nature sees it as a system ordered by physical and chemical laws. These laws are understood to operate everywhere in the universe at once, across vast expanses of matter. This idea motivated Andrew to assert at one SFI workshop that "nature is probably a parallel computation of some sort." If nature can be viewed as a computational system, it can easily be used to refer to processes in computers. At one SFI talk on computer simulation as a new tool for science, the speaker argued exactly this. He said that there is not just one nature but a variety of natures: a set of equations can be "a nature," and a computer experiment can be a nature. He wagered that there was a parallel between the process by which "Mother

Nature sets up laws" and the process by which a "computer experiment God sets up algorithms and initial conditions in 'Other natures.' "

In "An Evolutionary Approach to Synthetic Biology," Tom Ray writes that we must "understand and respect the *natural* form of the digital computer, to facilitate the process of evolution in generating forms that are adapted to the computational medium, and to let evolution find forms and processes that *naturally* exploit the possibilities inherent in the medium" (1994a:183; emphasis added; entire quotation italicized in the original). Ray says that Tierra can be considered an artificial world because it has its very own physics and chemistry—a " 'physics and chemistry' of the rules governing the manipulation of bits and bytes within the computer's memory and CPU" (1994a:184). Worlds are set up according to the rule of natural laws, and Tierra is no exception. A user may obtain a printout of a file called `soup_in1`, which lists the rules that govern Tierra. By changing these rules, a user can specify how frequent "mutations" will be, the way CPU time will be allocated to creatures, and so on (see fig. 6). Following the logic of Tierra as alternative nature, one supporter of Ray said at an SFI workshop, "I would argue that your work is empirical, not theoretical. You've built a new world and are doing empirical work in it."

For Ray, the future for artificial life looks bright once we recognize that computers can be full-blown worlds:

> Until recently, life has been known as a state of matter, particularly combinations of the elements carbon, hydrogen, oxygen, nitrogen, and smaller quantities of many others. However, recent work in the field of AL [Artificial Life] has shown that the natural evolutionary process can proceed with great efficacy in other media, such as the informational medium of the digital computer. These new natural evolutions in artificial media are beginning to explore the possibilities inherent in the "physics and chemistry" of those media. They are organizing themselves and constructing self-generating complex systems. While these new living systems are still so young that they remain in their primordial state, it appears that they have embarked on the same kind of journey taken by life on earth and presumably have the potential to evolve levels of complexity that could lead to sentient and eventually intelligent beings. (1994a:182–183)

I have been following Paul Edwards in understanding computers as closed worlds, worlds ordered and technically created as bounded conceptual spaces. Edwards writes that in ideology, in popular culture, and

```
# tierra core: 6-7-92

# observational parameters:

BrkupSiz = 1024       size of output file in K, named break.1, break.2 ...
CumGeneBnk = 1     Use cumulative gene files, or overwrite
debug = 0                     0 = off, 1 = on, printf statements for debugging
DiskOut = 1              output data to disk (1 = on, 0 = off)
GeneBnker = 1     turn genebanker on and off
GenebankPath = gb1/   path for genebanker output
hangup = 0                    0 = exit on error, 1 = hangup on error for debugging
Log = 1                 0 = no log file, 1 = write log file
MaxFreeBlocks = 800       initial number of structures for memory allocation
OutPath = td/   path for data output
RamBankSiz = 20000 array size for genotypes in ram, use with genebanker
SaveFreq = 2              frequency of saving core_out, soup_out and list
SavMinNum = 3            minimum number of individuals to save genotype
SavThrMem = .02   threshold memory occupancy to save genotype
SavThrPop = .02   threshold population proportion to save genotype
WatchExe = 0      mark executed instructions in genome in genebank
WatchMov = 0      set mov bits in genome in genebank
WatchTem = 0      set template bits in genome in genebank

# environmental variables:

alive = 2000      how many generations will we run
DistFreq = -.3        frequency of disturbance, factor of recovery time
DistProp = .2     proportion of population affected by distrubance
DivSameGen = 0 cells must produce offspring of same genotype, to stop evolution
DivSameSiz = 0 cells must produce offspring of same size, to stop size change
DropDead = 5 stop system if no reproduction in the last x million instructions
GenPerBkgMut = 8 mutation rate control by generations ("cosmic ray")
GenPerFlaw = 6       flaw control by generations
GenPerMovMut = 4     mutation rate control by generations (copy mutation)
IMapFile = opcode.map map of opcodes to instructions, file in GenebankPath
MalMode = 1 0 = first fit, 1 = better fit, 2 = random preference,
# 3 = near mother's address, 4 = near dx address, 5 = near top of stack address
MalReapTol = 1    0 = reap by queue, 1 = reap oldest creature within MalTol
MalTol = 20 multiple of avgsize to search for free block
MateProb = 0.0  probability of mating at each mal
MateSearchL = 5 multiple of avgsize to search 0 = no limit
MateSizeEp = 2 size epsilon for potential mate
MateXoverProp = 1.0 proportion of gene to secect for crossover point
MaxMalMult = 3      multiple of cell size allowed for mal()
MemModeFree = 0   read, write, execute protection for free memory
MemModeProt = 2 rwx protect mem: 1 bit = execute, 2 bit = write, 4 bit = read
MinCellSize = 12      minimum size for cells
MinTemplSize = 1      minimum size for templates
MovPropThrDiv = .5       minimum proportion of daughter cell filled by mov
new_soup = 1          1 = this a new soup, 0 = restarting an old run
NumCells = 2       number of creatures and gaps used to inoculate new soup
PhotonPow = 1.5       power for photon match slice size
PhotonWidth = 8    amount by which photons slide to find best fit
PhotonWord = chlorophill  word used to define photon
PutLimit = 10 distance for intercellular communication, mult of avg creat siz
ReapRndProp = .3 top prop of reaper que to reap from
SearchLimit = 5 distance for template matching, mult of avg creat siz
seed = 0 seed for random number generator, 0 uses time to set seed
SizDepSlice = 0  set slice size by size of creature
SlicePow = 1     set power for slice size, use when SizDepSlice = 1
SliceSize = 25     slice size when SizDepSlice = 0
SliceStyle = 2   choose style of determining slice size
SlicFixFrac = 0   fixed fraction of slice size
SlicRanFrac = 2   random fraction of slice size
SoupSize = 60000 size of soup in instructions
```

Figure 6.   Printout of the file soup_in1, displaying the rules that run Tierra.

in imagination, the alternative to the closed world has been not the open world but the "green world":

> The green world is an unbounded natural setting, such as a forest, meadow, or a glade. Action moves in an uninhibited flow between natural, urban, and other locations, and centers around magical, natural forces—mystical powers, animals, or natural cataclysms. . . . Green world drama thematizes the restoration of community and cosmic order through the transcendence of rationality, authority, convention, and technology. (1996:13)

In the imagination of Artificial Life scientists like Ray, the closed world comes together with the green world; computers are cultivated to contain a world open with possibility, resistant to total rational explanation, and full of magical (but still law-governed) evolutionary possibility. The title of the Artificial Life philosopher Claus Emmeche's book, *The Garden in the Machine* (1991), marks just this fusion of closed and green worlds.[6]

WESTERN CREATION STORIES

Definitions of the universe, worlds, or nature as law governed resonate with conceptions specific to Western Judeo-Christian[7] cosmology as it has been shaped in the wake of the scientific revolution; these definitions summon up images of a meticulous law-giving Creator and recall pictures of the universe as a giant clockwork or as a book in need of careful reading and deciphering. Many of the people I interviewed were raised as Jews or Christians, but virtually all now count themselves as atheists, a condition that did not prevent them when speaking of artificial worlds from relying on a host of Creationist mythologies, sometimes directly, sometimes channeled through the medium of science fiction. In one conversation, a researcher explicitly asked me "to think theologically for a moment" when trying to understand artificial worlds. He told me that there is a moment of creation when the programmer writes the formal rules that will govern the system. Andrew appealed to the possibility that our own universe might be just a cosmic simulation. He did this to buttress his arguments about why simulations might be considered worlds. He said, "If God up there turned off the simulator and then turned it back on again, we wouldn't know. So, that puts us in some kind of epistemologically inferior position, a lesser degree of reality than Him, since we can't do the reverse. And that would be true for the guys inside [the computer]."

Describing programmers as a genus of god was a frequent strategy among the people I talked with, and this was not just a playful way of speaking but a move that granted programmers the authority to erase their own presence as the beings who gave their simulations meanings as worlds. This move permits programmers to have the same relation to their simulations as the God of monotheism has to His creation (most vividly in his Deist incarnation as a divine watchmaker). Both occupy a transcendent position, the position of the "unmoved mover."[8] When I asked one researcher how he felt when he built simulations, he replied, "I feel like God. In fact, I am God to the universes I create. I am outside of the time/space in which those entities are embedded. I sustain their physics (through the use of the computer). I look like I am omniscient to the entities within that physics, and so on." Another researcher was pretty sure that simulations were divorced enough from our world to be considered self-contained worlds with their own dynamics:

> Except for our interactions—let's say we're sitting there and every once in a while we drop a blob of food in the thing, which apparently comes from God or something—I think [the simulation] is pretty much decoupled [from our universe]. From the point of view within that universe, it can have very different physical laws from our own, even though those laws are sort of implemented in terms of our laws. If, say, scientists evolved in that universe, they would come up with totally different theories for the laws of their universe.

Tom Ray holds that his "organisms" have their own purposes apart from his. He writes, "They were not expected to solve my problems, other than satisfying my passion to create life" (1992a:396). The SFI version of a program called the Echo system allows the user to "Choose," "Edit," "Run," "Pause," and "Replay" different worlds as if they were so many videotapes. In a clever denial of the ultimate control of the user, however, there is an error message in the program that reads: "You cannot set time backwards! Only God can do that." Bruce MacLennan's discussion of the experimental use of artificial worlds relies on images of a god who creates and then dispassionately observes the world: "Because synthetic ethology creates the worlds it studies, every variable is under the control of the investigator. Further, the speed of the computer allows evolution to be observed across thousands of generations; we may create worlds, observe their evolution, and destroy them at will" (1992:637). The use of this brand of god imagery allows Artificial Life researchers to disappear themselves from the scene and give their simu-

lations what the literary theorist N. Katherine Hayles has called "ontological closure" (1994a:4). One researcher told me over email, "AL researchers must write software to create their micro-world. This is standard control programming. Within the micro-world, however, the
evolving programs are out of control. So there is a mix there. We still
have to deal with tortuous programming in which every punctuation has
to be just right. Yet we are crafting an environment within which we intentionally give up control." This desire to push to one side the human
activity involved in interpreting artificial worlds was in evidence at the
MIT Artificial Life conference, when Chris Langton contended that
some simulations have less human agency embedded in them than others. The god imagery allows programmers to alternately shape and observe their words, to minimize the importance of their own postcreation
interventions, and to draw the boundaries between creation and tinkering strategically, so as to reinforce the sense that their systems might one
day have "free will."[9] The god imagery also betrays a desire for the scientist to become an omnipotent and omniscient god, to make this a practical reality in an artificial world that can be perfectly monitored and
controlled. As Ray put it in an interview with Kevin Kelly, "Even if my
world gets as complex as the real world, I'm god. I'm omniscient"
(quoted in Kelly 1994:297).

God imagery also inhabits descriptions of postcreation events, even
against the desire to keep it contained as merely a playful way to
talk about how artificial worlds come into being. Interventions after creation are routinely dubbed "miracles," in line with the definition of a
miracle as an outside intervention into the world that makes the impossible possible for a moment.[10] At one lecture on a simulated world, a
speaker discussed changing the conditions in a model while it was in
progress, and he said, "God can come down here and change some of
the os and 1s in the bit strings." Programming changes were figured as
divine intervention.[11] In another lecture, the philosopher Mark Bedau
showed a procession of graphs of activity in a simulation, and he explained a disruption recorded on one by saying, "I threw an asteroid into
the system." Here Bedau stepped into the role of a god capable of disturbing mundane reality. The figure of the asteroid also reinforced the
image of the computer as a world, even as a planet.

God talk is ubiquitous in popular treatments of Artificial Life. An ad-
vertisement for SimEarth, a home simulation of global ecological dynamics, reads: "More than just a home computer game, SimEarth was

developed with Professor James Lovelock, originator of the Gaia hypothesis, to give us all an opportunity to play god—from the safety and security of our own homes." Ad copy for SimLife, Ken Karakotsios's computer toy for experimenting with artificial evolution, reads: "Build your very own ecosystem from the ground up, and give life to creatures that defy the wildest of imaginations." During the time I was in Santa Fe, I played SimLife many times and was always struck by the process through which artificial worlds were created in the system. I would watch as a blank rectangular map—meant to suggest a god's-eye view—was progressively filled with "land," "mountains," and "water." After this creation, a smaller patch of place would be enlarged for my viewing and working convenience. A numbing number of pop-up menus would suddenly stand at attention alongside my square of world. I discovered that I could control temperature, rainfall, seasonality, humidity, mutagen level, length of days, and a bevy of other ecological parameters. Other menus accessed the menagerie of animals with which I could populate the world.

The theological luminescences that make SimLife thinkable are made clear in a review of the product written by Chris Langton:

> Maxis [the company that produces SimLife] provides the user with a fascinating virtual "nature," with its own physics and environment, replete with occupants that "live" their virtual lives within the confines of these artificial realities. The role of the user in these games is not so much participant in the action, as is the case with most computer games, but rather as the reigning "God" who designs the universe from the bottom up. . . . In SimLife, Maxis has essentially created a flight simulator that gives one a taste of what it would be like to be in the pilot's seat occupied by God. In fact, if God used a computer simulation to create the world and populate it with organisms, his software tools would look a lot like those found in SimLife. (1991:4)

Note the identification of God with a pilot in a flight simulator; this reminds us that the god we are supposed to have in mind is a god in heaven. This is a god who uses the tools of extraterrestrial technology to examine and create the world, and the satellitelike map provided in SimLife is a perfect tool for this transcendent, panoptic entity. This is a god in the image of the omnipotent scientist, a scientist already imagined as an omnipotent god. As Haraway notes, "Inside the still persistent Cartesian grid conventions of cyber-spatializations, the [Sim] games encourage their users to see themselves as scientists within narratives of exploration, creation, discovery, imagination, and intervention" (1997:132).

The theological resonances of artificial life worlds have not been missed by researchers themselves, though they have noticed these less as situating their science than as existing necessarily because they really are as gods. There is an Artificial Life bulletin board on the Internet called "alife.bbs.microcosmic_gods," which "is intended for discussion of Artificial Life as experimental theology." A mission statement:

> By creating artificial worlds, replete with creatures that are supposed to "live their lives" within them, we take on a "god-like" role with respect to these worlds and their inhabitants. As a result, as a matter of practical day-to-day experience, we encounter many of the issues that commonly fall into the domains of theology and religion (one simulation system even provides a "smite" option, allowing the user to go around blasting creatures out of virtual existence with a little lightning bolt). How can we use Artificial Life models as vehicles to enhance our understanding of some of these theological issues?

This statement ignores the fact that the theology it speaks of is fully in line with a Judeo-Christian imaginary. In a discipline concerned with *life-as-it-could-be,* it is interesting that only one version of theology is entertained, only one imagining of "god" is in place. God is understood to be transcendent, all-powerful, and occasionally malicious, much like the God of the Old Testament. It is also understood that any world has only one god at a time. The playful character of this newsgroup belies the serious way some Artificial Life scientists have secularized very traditional religious questions. Even as atheists or agnostics, they have not abandoned the traditional paradoxes of Western theology. In fact, we might say that these enable thinking of Artificial Life in a deep way. We are never given a creation in which the programmer is akin to Vishnu dreaming up a new reality, or part of a polytheistic team of gods tinkering and competing in constructing a world they are never completely outside—though this last possibility is suggested by the collaborative aspects of some systems, and at least one researcher has attempted to fuse Artificial Life with Multi-User Domains, pursuing the idea that distributed human users might like to shape a space shared with artificial life-forms on the Internet (see Ackley 1997). Of course, the most Christian issues in theology never really come up either; programmers do not imagine downloading themselves into their simulations as virtual Christs, visiting their creatures in an embodied, personal way and offering them salvation.

While Artificial Life researchers might deny the constitutive role of creation language in making their simulations, this language has actually been potent enough to convince some Creationists that Artificial Life systems in fact prove that life always needs an author. This is ironic, since one of the particular crusades of Murray Gell-Mann of SFI is the refutation of Creationism (Gell-Mann 1994), and he had hopes that computer models of evolution could convince people that God need not exist for complex and diverse species to come into existence.

SCIENCE FICTION

Science fiction was an extraordinarily rich resource for scientists' imaginings of computers as worlds. Many researchers have been avid science fiction fans and have self-consciously used the genre as a launching pad for scientific speculations. This is most true for computer scientists who came of age during or after the 1960s; older scientists at SFI have little familiarity with science fiction and, as I have noted, short patience with the notion that computers are universes. Several people in the 1960s cohort referred me to stories in which artificial worlds were created by entities who imagined themselves gods. A few told me of Theodore Sturgeon's "Microcosmic God" (1941), in which a scientist creates a brood of quickly evolving organisms on which he practices eugenics and to whom he issues command(ment)s via teletype. Larry Yaeger said that he was so taken by the story that he printed "Microcosmic God" on his business card. The Polish science fiction author Stanislaw Lem was also a frequent reference, particularly his story in *The Cyberiad* (1967) in which Trurl the constructor (himself a robot) builds a miniature kingdom made of mathematics. In this tale, penned as a political critique of instrumental rationality, Lem discourses on the difficulty of telling mathematical formulations apart from reality. I reproduce a fragment of the story here, for it is almost a scriptural piece for Artificial Life. Note how the discussion between the two characters anticipates almost exactly the conversation between Chris Langton and Stevan Harnad—and anticipates it so exactly that one has to conclude that science fiction informs science as much as the reverse.

Trurl the constructor has just returned home to explain to his friend Klapaucius that he has fashioned a model of a kingdom for a despotic king. He feels that he has given this king a toy that will keep him out of trouble, and keep him from oppressing real creatures.

"Have I understood you correctly? [Klapaucius] said at last. "You gave that brutal despot, that born slave master, that slavering sadist of a pain-monger, you gave him a whole civilization to rule and have dominion over forever? . . . Trurl how could you have done such a thing?!"

"You must be joking!" Trurl exclaimed. "Really, the whole kingdom fits into a box three feet by two by two and a half . . . it's only a model . . . "

"A model of what?"

"What do you mean of what? Of a civilization, obviously, except that it's a hundred million times smaller."

"And how do you know there aren't civilizations a hundred million times larger than our own? And if there were, would ours then be a model? And what importance do dimensions have anyway? In that box kingdom, doesn't a journey from the capital to one of the corners take months—for those inhabitants? And don't they suffer, don't they know the burden of labor, don't they die?"

"Now wait just a minute, you know yourself that all these processes take place only because I programmed them, and so they aren't genuine. . . . "

"Aren't genuine? You mean to say the box is empty, and the parades, tortures and beheadings are merely an illusion?"

"Not an illusion, no, since they have reality, though purely as certain microscopic phenomena, which I produced by manipulating atoms," said Trurl. "The point is, these births, loves, acts of heroism and denunciations are nothing but the minuscule capering of electrons in space, precisely arranged by the skill of my nonlinear craft, which—"

"Enough of your boasting, not another word!" Klapaucius snapped. "Are these processes self-organizing or not?"

"Of course they are!"

"And they occur among infinitesimal clouds of electrical charge?"

"You know they do."

"And the phenomenological events of dawns, sunsets and bloody battles are generated by the concatenation of real variables?"

"Certainly."

"And are we not as well, if you examine us physically, mechanistically, statistically and meticulously, nothing but the minuscule capering of electron clouds? Positive and negative charges arranged in space? And is our existence not the result of subatomic collisions and the interplay of particles, though we ourselves perceive those molecular cartwheels as fear, longing, or meditation? And when you daydream, what transpires within your brain but the binary algebra of connecting and disconnecting circuits, the continual meandering of electrons?"

Lem 1967:166–167

Trurl is being accused of playing God, a god that is a creator, a unitary entity, and fully outside creation. In this story, and in Sturgeon's, we reread a familiar theology, translated into evolutionary and cybernetic jargon. The Judeo-Christian cast of the ways Artificial Life researchers talk about their "worlds" comes to them not because they participate in these religious beliefs (they don't, for the most part) but because the resources they have at hand for imagining alternative worlds, resources like science fiction, come to them heavily dosed with Bible stories.

Science fiction culture and common sense encodes a number of cultural themes aside from those having to do with creation. The most salient is the theme of exploring and colonizing the universe, a theme in many ways a rewriting of American frontier and cowboy tales. Many researchers told me that creating artificial worlds in computers is necessary for a universal biology since, at the moment, we are unable to do natural history on other planets. Since we do not have *Star Trek*'s USS *Enterprise* at our disposal, we must somehow construct the worlds we would explore. This science fictional imperative to canvass the universe is given a theoretical warrant in the official ideology of Artificial Life. As Chris Langton puts it in a promotional video for SFI: "It's very difficult to build general theories about what life would be like anywhere in the universe and whatever it was made out of, when all we have to study is the unique example of life that exists here on Earth. So, what we have to do—perhaps—is the next best thing, which is to create far simpler systems in our computers" (Langton in Santa Fe Institute 1993). If we found life on another planet, Langton frequently says, we would not want to exclude it from our science of biology. Doyle has commented that the idea that theoretical biology is somehow "hamstrung" by its imprisonment on earth derives from a desire to occupy a transcendent position from which to scan the universe (1997b:111–112). It is precisely this position that Artificial Life scientists strive to create for themselves by manufacturing their own galaxies of artificial worlds.

This notion of a universal biology, underwritten by a theoretical physics faith that there must be such a thing, depends crucially on a science fiction imagination. One young man defended Langton's intuition that life is a formal process that can inhabit different sorts of matter:

> I think Chris's statement has to be true, because if we someday went out traveling through space and found some other life-forms that were made

out of something entirely different, we wouldn't say they're not alive, just because they don't have carbon in them. If Chris's statement weren't true, then we might as well just give up, but to me that statement is intuitively obvious. It seems ridiculous to me to assume that this is the only way that life could be built, but I read a lot science fiction as a kid, so . . .

When I asked one researcher how he would explain his research to a child, he said he would start by asking his listener to imagine life on other planets, in other chemistries, in other physical substrates. Then, he would escort the person back to our planet and ask her or him to consider the possibility of other life-forms coming into being right here at home, maybe even inside the physical universe of computers. I mentioned that Darwin had used kinship metaphors to persuade his readership of the evolutionary connectedness of all life on earth. He said that he was making a similar move by asking people to imagine life on other planets before considering Artificial Life on earth. He was asking people to generalize from a phenomenon with which he assumed they were already familiar.[12]

We might say that Artificial Life researchers are frustrated that they are not living in the future. Like people who have their bodies or heads put in cryonic suspension, they are impatient for the future to arrive, impatient enough to want to reel it into the present (Doyle 1996, 1997a). One person said in conversation with me that he was frustrated by his own finitude and mortality. He wished he could live long enough to see "all the cool things that would happen in the future." He wished he could be around when humans eventually contacted intelligent life on other planets, and his desire to create life in computers had to do with this curiosity. In this vision, cyberspace is figured as the new outer space, the cosmos that the various space programs have failed to deliver us to.

Perhaps the impulse to think of computers as worlds is served in part by a desire to author a reality, just as a science fiction writer does. Andrew told me that he thought of himself as a storyteller and that programming, like the Dungeons and Dragons designing he used to do, was one way of telling stories. At a cocktail party at which Andrew was one center of attention, his nonscientist conversation mates said that from the way he described his work, he was more of an artist than a scientist. Le Guin's essay "Do-It-Yourself Cosmology" provides a useful account of how science fiction writers create speculative alternative worlds. Much of this speaks directly to the ethos behind Artificial Life. Le Guin writes of an article by Poul Anderson called "How to Create a World":

Mr. Anderson provided a good batch of facts the universe maker wants, including several mathematical equations useful in various situations. . . . This kind of world-making is a thought-experiment, performed with the caution and in the controlled, receptive spirit of experiment. Scientist and science-fictioneer invent worlds in order to reflect and so to clarify, perhaps to glorify, the "real world," the objective Creation. The more closely their work resembles and so illuminates the solidity, complexity, amazingness and coherence of the original, the happier they are. (1989:119–120)

This is manifestly clear in Chris Langton's review of SimLife:

SimLife allows the construction and study of simple, artificial ecologies, and the surprising complexity and richness of behavior that emerges in even these extremely simple "artificial natures" has already given me a greater appreciation of the real thing—Nature in all her real glory—writ with a capital "N." That glory shines all the more brightly even from the meagre illumination that this simple "Software Toy" is able to shed upon Her. (Langton 1991:6)

Science fiction-fueled simulation has become a new tool for thinking natural theology, a new tool for revealing and reproducing the plans of (a) God.

Of course, there are a variety of science fiction literatures, and Artificial Life researchers mostly refer to a fairly traditional sort, concerned with robots and space (and now, cyberspace) colonization. Feminist and queer science fictions (like those of Ursula Le Guin, Octavia Butler, Samuel Delany, Marge Piercy, and Joanna Russ) also imagine alternative worlds, but they capture a very different sort of imagination, one more about undermining dominant views of gender, sexuality, species, and life than about transporting received stories into new hardware.[13]

CYBERSPACE AS A NEW CREATION AND COLONIAL SPACE

The idea that computational processes can be thought of as existing in a kind of territory is supported by popular talk about computer networks as "cyberspace." The historian of science David Noble summarizes the often millenarian, Christian language used in discussing cyberspace:

The religious rapture of cyberspace was perhaps best conveyed by Michael Benedikt, president of Mental Tech, Inc., a software-design company in Austin, Texas. Editor of an influential anthology on cyberspace, Benedikt argued that cyberspace is the electronic equivalent of the imagined spiritual realms of religion. The "almost irrational enthusiasm" for virtual reality, he observed, fulfills the need "to dwell empowered or enlightened on other, mythic, planes." Religions are fueled by the "resentment we feel for our bodies' cloddishness, limitations, and final treachery, their mortality. Reality is death. If only we could, we would wander the earth and never leave home; we would enjoy triumphs without risks and eat of the Tree and not be punished, consort daily with angels, enter heaven now and not die." Cyberspace, wrote Benedikt, is the dimension where "floats the image of the Heavenly City, the New Jerusalem of the Book of Revelation. Like a bejeweled, weightless palace it comes out of heaven itself . . . a place where we might re-enter God's graces . . . laid out like a beautiful equation." (1997:159–160)

Computers are figured as places to begin again. Artificial Life participates in this imaginary by maintaining that simulated worlds might be places to see other ways life could have evolved. Ray's Tierra is a freshly minted artificial world that has yet to know anything like the Cambrian period's eruption of diversity. At one conference, I saw a simulation called "Aleph," a nice pun on ALife, but also a name that summons up, with its Hebrew letter name, the beginning of things. *The Quest for a New Creation,* the subtitle of Steven Levy's popular book, *Artificial Life* (1992b), captures the imagination at work here.

This image of computers as new worlds goes well with another image popular these days: the "electronic frontier." As Robert Bellah (1992) and many others (e.g., Obenzinger 1994) have argued, notions of the holy land and notions of the frontier (particularly the American frontier) are often crafted together. People imagine new frontiers as allowing them to start anew from an Edenic state: early English colonists saw in America a land of milk and honey, the pilgrim John Winthrop saw in America a "City upon a Hill" (a reference to a new Jerusalem), and Mormons saw in Utah a new Zion. The Americas, in their disguise as "the New World," were often understood as a place for (European) humanity to begin again. In the frontier territories of Artificial Life, we find that "Man" will be as God again, having regained his creative capacity modeled after God. He will create creatures and give them their rightful names, just as Adam did, and just as Carolus Linnaeus—the eighteenth-century taxonomist who called himself a "second Adam"—did at the moment modern biology was born. The image of a Garden of Eden in cyberspace dovetails with the spatialized metaphor of the electronic frontier and allows Artificial Life

researchers to think of themselves as creating, populating, and exploring new lands.

The idea that computer networks can be thought of as "cyberspace"
trades on a territorial metaphor that is profoundly American. The
popular press tells us that cyberspace is full of outlaws, that it is an an-
archic space much like the Old West. We are told that cyberspace will
soon become a rational democratic public space, an information superhighway on which we can travel where we please. This notion of colonizing a new space is prominent in many Artificial Life narratives. One
researcher told me that examining the different possibilities inherent in
the simple rules he programmed into his system was like "exploring a
new land." This notion of colonization and taming underwrites Tom
Ray's plan to have a global version of Tierra (see Ray 1994b). To hear
Ray speak of Network Tierra is to hear him speak of an empty world he
plans to colonize with his digital organisms. Cyberspace, like the ever-
out-of-reach outer space, becomes a new frontier, a silicon second nature ready for colonization by the first world imagination.[14] Ray's project is impossible to conceive without the territorial metaphor, without
the colonial imagination and its Creationist underpinnings.[15]

In a paper entitled "Visible Characteristics of Living Systems," published in the proceedings of the second European Conference on Artificial Life, a company of philosophers and artists clearly articulate the idea
that Artificial Life researchers think of themselves as exploring and colonizing new spaces:

> Artificial Life researchers habitually give names to their creations, build up
> complicated typologies for them and come up with bold new syntactic des-
> ignations for areas that are still terra incognita. They are the true successors
> of the Conquistadors of the 15th and 16th centuries, of the explorers of the
> 18th and 19th. . . . Having created his creature in real or virtual space, the
> researcher or artist-researcher tries to capture it with language and to mas-
> ter its imaginative dimension in such a way as to turn it into a usable ob-
> ject, one that can be proposed for comparison, criticism, and reconstruc-
> tion. Putting AL researchers on the same plane as the Conquistadors is not,
> then, a cheap provocation, for each draws his inspiration from a voyage
> into the unknown. (Lestel, Bec, and Lemoigne 1993:598)

The authors almost celebrate this image, leaving little room for seeing
how Artificial Life is not just analogous to projects of naming and conquest but fundamentally informed by them. But perhaps my interpretation of Lestel and friends is too hasty, for they continue:

A major difference clearly separates the Conquistadors from the AL research-ers of the late 20th century. The former were crossing a pre-existing region, a land that had lived without them for centuries and which was endowed with an extremely rich life of its own. The situation of AL researchers is different and more ambiguous, for they are exploring intellectual and technological territories that they have to create as they go along. Yet the Conquistadors also saw themselves as exploring a land that was very largely that of their dreams, hopes, fantasies, and long-held desires; thus they developed a rela-tionship with the past akin to that of AL researchers, who draw on an extra-ordinarily rich cultural, intellectual and technological past which no-one has yet exploited to the full and of which they are equally inheritors and creators. (Lestel, Bec, and Lemoigne 1993:599)

In an email interview, SimLife inventor Ken Karakotsios speculated that Artificial Life might allow people to manufacture and pioneer new civilizations in computers: "I'm most interested in exploring ALife as a new frontier, a new medium for the expression of life and possibly new cultures. I don't think we're ever going to make interstellar travel prac-tical, so I hold little hope of extensive exo-cultural exchange. However, I think there are an infinite number of very worthwhile cultures that will be created in simulated universes." It is not clear what Karakotsios means by culture here (let alone by "worthwhile cultures"), but I would like to pull out the meaning of "culture" as the product of the cultiva-tion of living material in a prepared nutrient medium.

The rhetorics of exploration, colonization, and conquest in Artificial Life are intensely masculine. Imaginings of programmers as gods, as single-handed creators of life, as transcendent observers, and as intrepid explorers of the final frontiers of cyberspace all invoke phenomenally masculine imagery—and not just this, but imagery of a kind of white masculinity, of a white man who hunts, explores, and goes on adven-tures in undiscovered lands and feels assured in his power in naming and conquering. That many Artificial Life practitioners are white men who grew up reading cowboy science fiction is not trivial. The location of SFI in New Mexico, a place associated with the days of westward frontier expansion, is also fitting, and acts as a resource for imagery enabling the crafting of computers as worlds.[16] The masculine imagery of exploration also produces a gendering of the "lands" of cyberspace as feminine, as waiting to be penetrated or unveiled, continuing a Western tradition of seeing Nature as a woman (recall Langton's pronouncements about "Na-ture in all her real glory" above.) Nature, a living—sometimes nurturing,

sometimes wild—feminine being in seventeenth-century European cosmology was devivified and mechanized in the eighteenth and nineteenth
centuries (Merchant 1980), and has now been resuscitated as an enormous computer.

PSYCHEDELIA

Though my interviewees owed their first epistemological allegiances
to science, many middle-aged and younger researchers told me at parties and over informal lunches of quasi-mystical epiphanies that had
helped them see the natural world in new ways. A few told me of early
experiences with psychedelic drugs that allowed them to understand
parts of the inanimate world as infused with life, a perception
perhaps not available within the usual categories of Euro-American
thought (but one amenable to rephrasing in scientific terms as the notion that life is a continuum process across grades of organizational complexity). Some intimated that LSD had awakened them to the possibility that the world is what we make it. One said, "A lot of my experience
with LSD has made me feel very strongly for a view of the world as a
fiction created by our own biology." These experiences, species of
counterculture events domesticated into tales of scientific inspiration,
alerted people to the idea that there might be many alternative worlds,
many separate realities.

I was set on the trail of the psychedelic influence on Artificial Life by
Gerald, who recommended I take seriously this connection. He wrote
me over email,

> One aspect of ALife that might be worth considering is the role that psy-
> chedelic drugs play in the imagination of ALife researchers and in the inter-
> pretation of results. The ALife community has more of a psychedelic
> sensibility than a metaphysical or religious sensibility. Roughly, the follow-
> ing analogies could be made:
>   biblical scripture = research findings;
>   spiritual visions = psychedelic/subconscious interpretations of ALife
>   concepts;
>   religious revival meetings = science conferences, operating at the inter-
>   section between objective findings (scripture) and their psychedelic
>   interpretation and elaboration (visions).
> If ALife did not have this sort of psychedelic dimension, it would not at-
> tract nearly so much interest or enthusiasm. It would be like scripture with-
> out prayer or hallucination.

Gerald was certainly on to something; as I conducted my interviews, I noticed that there were a few people who explicitly tied their experiences with mind-altering drugs to near-spiritual experiences they had watching the results of their first computer simulations.[17] Stories of the foundation of Artificial Life recount the epiphanies of central scientists. We learn in various places that Chris Langton once felt spooked by a Game of Life running on a screen in a hospital where he was working, that he felt as he looked away from the monitor as though something was in the room with him. Doyle has written about this moment and others like it as instances of a "glance away" that brings Artificial Life simulations to life for researchers, a moment in which fleeting vision endows everyday items with an eerie vitality (1997b:126). This brings me to a discussion of the significance of technologies and rhetorics of visualization in making worlds out of computers.

TECHNOLOGIES OF VISION

The data researchers garner from their simulations are usually presented visually, and in Artificial Life simulations, the visuals almost always afford a god's-eye view, allowing experimenters to survey an entire world at once. In PolyWorld (see fig. 3) and SimLife (fig. 7), we are provided with a panoptic picture of a world, over which we can witness the careenings of computational creatures. In Tierra and programs like it, we are presented with graphs that summarize evolutionary activity and chart the rise and fall of program lineages. The visual access we are granted to artificial worlds positions us as observers located everywhere and nowhere at once. Artificial Life programs instantiate the dream of objectivist science. Haraway has written that in Western thought, "the eyes have been used to signify a perverse capacity—honed to perfection in the history of science tied to militarism, capitalism, colonialism, and male supremacy—to distance the knowing subject from everybody and everything in the interests of unfettered power. . . . [The] view of infinite vision is an illusion, a god-trick" (1991f:188–189). This illusion is manufactured as reality in computational Artificial Life.[18]

Vision was a sense researchers often invoked to speak about untheorized and unmediated pictures of the empirical world. They spoke of "seeing" phenomena emerge in their simulations and when speaking this way, took themselves to be reporting—rather than, say, hallucinating—real results. At many conferences I attended, people referred to the "real

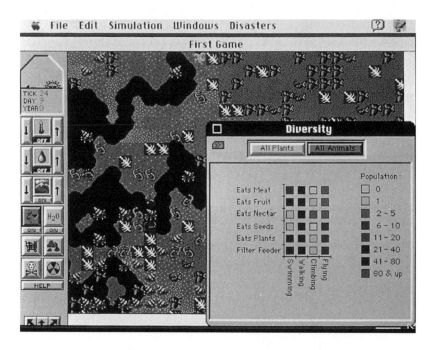

Figure 7.   Screenshot of SimLife. Copyright 1992 Ken Karakotsios.
Reproduced with permission.

world" or "nature" by pointing out the window while they indicated artificial worlds by pointing at images on computer screens. At an SFI
workshop on artificial worlds, one researcher compared the technology
of computer simulation to telescopes and microscopes, saying that these
technologies allow us to peer into whole new worlds.[19]

Television is an important reference point for understanding how to
look at artificial worlds. Not only are these worlds beheld on exactly
the same kinds of screens as television programs (and video games, for
that matter), but television viewing habits are assumed by the technologies. Worlds begin when machines are turned on, when light flick-
ers forth from the computer screen. And, as with television, one can become addicted to artificial worlds. Chris Langton said in a lecture, "You
can spend a lot of time generating these images and forget that there's
another world out there." In related imagery, the world is frequently
compared to a videotape, one that we might rewind, fiddle with, and
watch repeatedly for different story lines. Langton writes, "Although
studying computer models of evolution is not the same as studying the

'real thing,' the ability to freely manipulate computer experiments, to 'rewind the tape,' perturb the initial conditions, and so forth, can more than make up for their 'lack' of reality" (1992:7–8; see also Fontana and Buss 1994).[20]

This image of the world as videotape fast-forwards me to a description of a compelling film I saw at the MIT Artificial Life conference. During one of the first talks, "Artificial Fishes with Autonomous Locomotion, Perception, Behavior, and Learning in a Simulated World," the speaker showed a video of stunning simulations of fish locomotion (see Terzopoulos, Tu, and Grzeszczuk, 1994; and fig. 8). The audience was enraptured as simulated fishes acted out their artificially evolved capacities to swim and hunt. The presentation ended with some of the simulations strung together in an extended (and sound-tracked) parody of a Jacques Cousteau nature documentary. As the audience laughed at the video, and at the movements of the fishes on the screen, it became clear that the "lifelike" quality of these simulations produced an unease and sense of wonder that was itself precisely the cultural resource that made these creatures seem lively. The laughter bespoke a set of unarticulated intuitions and untheorized thoughts about autonomy and agency.

The reference to Cousteau evokes another ocular technology: the aquarium, which also interposes a glass barrier between an observer and a bundle of phenomena. The aquarium has fashioned for us a way of manufacturing, thinking about, and seeing self-contained ecologies. It is an exhibitionary technology that creates a closed world and that creates a sense that worlds are just bigger aquariums or terraria. A few Artificial Life researchers call on people's familiarity with aquariums or terraria to rhetorically structure their simulations as small worlds. Some systems show us fishes or birds moving around as though they were in an enclosed space. Aqua Zone is an Artificial Life toy that invites us to treat our computer as a fish tank (we even feed the artificial fishes through the disk drive!). Jakob Skipper's article in the proceedings of the first European Conference on Artificial Life is called "The Computer Zoo—Evolution in a Box" (1992). The artificial world has a lineage in practices of exhibiting living creatures, as I was reminded when I visited the Sea Life Centre in Brighton during a day off from a conference on the simulation of adaptive behavior. The effect of looking at creatures through glass, especially creatures that moved in ways similar to some Artificial Life creatures, was uncanny. I saw many scientists from the conference at the Centre, traveling from one aquarium to another,

Figure 8.   Artificial fishes from Terzopoulos, Tu, and
Grzeszczuk 1994. Reproduced with permission.

contemplating the familiar and the unfamiliar life gleaming and gliding behind glass.

All of this viewing of "worlds" and "life" behind glass suggests still another set of processes through which visualization technologies produce "life": practices of imaging blastocysts, zygotes, embryos, and fetuses inside women's bodies through technologies such as electron microscopy and ultrasound (see Duden 1993; Haraway 1997). What is remarkable about these imagings is that they assume the existence of the entities they purport to represent, and so in a very material sense, end up *producing* these things as real. They isolate and materialize a zygote or fetus as an already existing individual, or even as "a life" (Duden 1993). Images of artificial life-forms on computer screens accomplish the vivification of entities through relying on a similar activity of visualizing entities presumed to be real but invisible. Here are entities that some say are hidden from the spectator's view because they ultimately exist as patterns of voltage in the computer but that can be rendered visible as dancing patterns on a cathode ray tube screen. We might say of artificial life organisms what Haraway says for images of the fetus, "The visual image of the fetus is like the DNA double helix—not just a signifier of life, but also offered as the-thing-in-itself. The visual fetus, like the gene, is a technoscientific sacrament. The sign becomes the thing itself in ordinary magico-secular transubstantiation" (1997:178).

A final comment on artificial worlds and vision. Hayles has argued that U.S. national parks have trained white-collar people in understanding nature as something to be grasped visually, a fact that she sees as having an interesting consequence:

> When "nature" becomes a object for *visual* consumption, to be appreciated by the connoisseur's eye sweeping over an expanse of landscape, there is a good chance it has already left the realm of firsthand experience and entered the category of constructed experience that we can appropriately call simulation. Ironically, then, many of the experiences that contemporary Americans most readily identify with nature—mountain views seen from conveniently located lookouts, graded trails traversed along gurgling streams, great national parks like Yosemite visited with reservations made months in advance—could equally well be considered simulation. (1995:411)

Like artificial worlds, national parks are often viewed through a glass barrier, in this case that sheet of curved glass known as the car window.

THE ARTIFICIAL WORLD WORKSHOP

The view of computers as artificial worlds is fundamental enough to SFI science that, in November 1993, a workshop was held to hash out what it might mean for scientific practice and theory. Organized by Walter Fontana, a chemist, and David Lane, a statistician, the workshop's primary purpose was to close in on why computers might be worlds in most senses that matter. It was also aimed at figuring out how people from different disciplines might usefully collaborate in computer modeling. Present at the workshop were economists, biologists, political scientists, people from business schools, and one man from the World Bank. All were hopeful that the workshop would result in clear conclusions about how to interpret computer simulations. Fontana and Lane were most concerned with epistemology, and because they had a strong sense that it was language that allowed processes in the computer to be mapped onto processes in the world, they invited the linguist George Lakoff, who has written extensively on metaphor (see particularly Lakoff and Johnson 1980).

The workshop was introduced by Lane, who outlined what artificial worlds allowed as far as he was concerned. He said, "Artificial worlds are computer-implementable stochastic models, which consist of a set of 'microlevel entities' that interact with each other and an 'environment' in prescribed ways. Artificial worlds are designed so that they themselves may, under some conditions, manifest emergent hierarchical order." Lane argued that the key questions in artificial worlds are how agents should be represented, at what level of abstraction they should work, and how we might make inferences about their interactions. It was apparent to him that all the artificial worlds under construction at Santa Fe had something in common and that SFI would be well served to figure out what that was. Lane's notion of world did not begin at physics but rather at what might be deemed a "higher level," a level at which entities interact in ways that construct the realities in which they participate and that in turn structure the rules governing interaction. The artificial worlds Lane spoke of were programmed in object-oriented programming languages (like C++), languages that allow the programmer to construct a library of computational entities that can interact in multiple ways and contexts, primarily through sending and receiving messages.

Fontana spoke next, centering his comments on the place of language in constructing worlds. He said that computational modeling of phenomena above the level of physics can be accomplished when we understand that computers are in their essence formal languages and that worlds can be understood to be made of language. He said, "By 'world' in the label 'Artificial World,' I mean a world which has language as an ingredient, as an essential component." He argued that language is a device that forces attention to construction, the genesis of organization, and the building of entities. Language "specifies an infinite universe with finite means, and therefore emphasizes the process of how to obtain the elements of this universe, as opposed to an extensional description where the universe is assumed to be available instantaneously in its entirety." Because the computer is in its essence a formal language, it is based on construction, on building entities. A computer is a language because software is a language and there is no mathematical difference between hardware and software. A computer is thus a tool for expressing or realizing artificial worlds. Fontana put it more strongly in an email interview: "Computers are candidates for *being* worlds, not just describing them. . . . The computer is not just a tool for fast calculations, sequence analysis, database management, etc. The computer is a medium." It is a medium, like the world, in which entities can be constructed out of the logics of low-level systems like chemistry. The formal and linguistic definition of a universe allows people to take literally the idea that a universe is an abstract entity that might be implemented in different media. One person drove this home to me when he said of evolving computer programs, "You really have to consider the process alive because you could transfer it to a different computer and it wouldn't know essentially, unless there were some type of physical sensors on the thing."

If, as many people at this workshop held, language is a tool for representation as well as a tool for making worlds, then representation and reality collapse into one another. Computer simulations are maps that become territories. As Steven Levy puts this view, "Although, indeed, a computer experiment is by no means equivalent to something it may be modeled on, it certainly is *something*. Maps are not *the* territory, but maps are indeed territories" (1992b:337). Descriptions, texts, and codes can become worlds. The philosopher Jean Baudrillard, in his vertiginous discussion of simulation, hints at the logic that has made artificial worlds real worlds:

> Simulation is no longer that of a territory, a referential being or a substance. It is the generation by models of a real without origin or reality: a hyperreal. The territory no longer precedes the map, nor survives it.

Henceforth, it is the map that precedes the territory—PRECESSION OF
SIMULACRA—it is the map that engenders the territory. . . . But it is no
longer a question of either maps or territory. Something has disappeared:
the sovereign difference between them that was the abstraction's charm.
(1983:2)

At the end of the workshop, Lakoff spoke of how he had heard participants using language and metaphor in describing these "worlds." He
pointed out that using logic from one domain to discuss another allows
one to import the inferential structure used to reason about the first
domain into the second. Understanding computer simulations "empirically" is guided by conceptual systems, and there can be no way to separate our language from their reality; they do not exist objectively, apart
from us. But while SFI and practitioners of Artificial Life may come up
with a stable practice for interpreting computer simulations as model
worlds, or even as real, independent worlds, this does not mean that they
become such things for people who do not share the enabling metaphors.
Lakoff maintained that what models do is "create an expert priesthood,
and what SFI is doing is setting up a new kind of priesthood." The anthropologist Lars Risan has also suggested this and has argued that just
as the authority of Catholic priests empowers them to convince people
that wine can be literally, not just metaphorically, turned into blood, so
it might be that the expert priesthood of Artificial Life, through skilled
interpretation of computer simulation, can convince people that computers and computer simulation can be literally turned into worlds harboring life (Risan, personal communication). The sociologist of science
Bruno Latour has declared that scientific facts only exist outside the laboratory to the extent that scientists transform pockets of the world into
corners of the laboratory. As he puts it, "Scientific facts are like trains,
they do not work off their rails" (1983:155). Artificial Life rhetorics have
already made their way into everyday common sense through such programs as SimLife, extending the network of priestly training in a kind of
Protestant dispersal of Artificial Life (see Doyle 1993).

Simulation also acts as a powerful medium for the transport of ideas
from one discipline and one lifeworld to another. People translate from
one language to another through the common currency of computer
simulation. In the process, the ontologies of the world all have to be
steamed down to entities that can be commonly represented as information structures, and this process often ends up consolidating the idea
that there really is something "deep" that diverse systems have in com-

mon. Peter Galison (1996, 1997) has spoken of the "trading zone" created in such interactions and has written about computer simulation as
a site where languages from different disciplines are coordinated. But
whereas Galison argues that this mutual calibration of language does not
necessarily commit scientists to a common ontology, a common view of
the fundamental constituents of reality, I would say that, at least in the
context of SFI, it does: an ontology of information.

Lakoff argued that metaphorical armatures allow for formal
processes of computation to be mapped onto words with both scientific
and commonsense meanings. I follow this argument at various points in
this book, maintaining that as long as science is done by creatures who
use language, there can be no space free of metaphor. The point is not
to eliminate metaphor or to discipline language into transparency but to
attend to how language is used and to the meanings these usages summon into the sciences. As the historian Evelyn Fox Keller and the philosopher Elisabeth A. Lloyd write about the connections between scientific
and social acceptations of words, "By virtue of their dependence on ordinary language counterparts, technical terms carry, along with their ties
to the natural world of inanimate and animate objects, indissoluble ties
to the social world of ordinary language speakers. . . . [They] have insidious ways of traversing the boundaries of particular theories, of historical periods, and of disciplines. . . . They serve as conduits for unacknowledged, unbidden, and often unwelcome traffic between worlds"
(1992:1–2).

Once one has accepted that simulations can be reasonable models of
worlds, then many words used to describe things in the world—like "living things"—can be extended to objects in the computational medium.
In the looking-glass worlds of Artificial Life, researchers see a permutation of the world with which they are familiar and extend their culturally constituted intuitions to fabricating and interpreting phenomena
that emerge *in silico*. In the next segment of this book, I argue that these
looking-glass worlds reflect many of the assumptions and lived experiences of researchers about what constitute creation, life, primitivity, individuality, kinship, and economics. I am particularly keen to track the
"unacknowledged, unbidden, and often unwelcome traffic" between the
social worlds and artificial worlds of Artificial Life.

THREE

Now, if you'll only attend, Kitty, and not talk so much, I'll tell you all my ideas about Looking-Glass House. First, there's the room you can see through the glass—that's just the same as our drawing-room, only the things go the other way. I can see all of it when I get upon a chair—all but the bit just behind the fireplace. Oh! I do so wish I could see *that* bit! I want so much to know whether they've a fire in the winter: you never *can* tell, you know, unless our fire smokes, and then smoke comes up in that room too—but that may be only a pretence, just to make it look as if they had a fire. . . . Oh, Kitty, how nice it would be if we could only get through into Looking-Glass House! I'm sure it's got, oh! such beautiful things in it! Let's pretend there's a way of getting through into it, somehow, Kitty. Let's pretend the glass has got all soft like gauze, so that we can get through. Why, it's turning into a sort of mist now, I declare!

LEWIS CARROLL, *Through the Looking-Glass*

Inside and Outside
the Looking-Glass Worlds
of Artificial Life

THIS CHAPTER IS organized as an extended tour of some artificial life worlds. I want you to imagine yourself at an Artificial Life conference, moving from one simulation to the next, as you might stroll from one tank to another at a sea life museum. I have set up a sequence of exhibits, and my choice and ordering of these digital dioramas is far from innocent; there are points I want to make, and I have chosen specific simulations to do so. Artificial Life worlds almost invariably begin with a creation story, and scientists' reports on their experimental and observational findings in these spaces often resemble natural historical narratives. They take us from tales about the origins and essence of life to stories of primitive organisms and to chronicles of the rise and fall of families, lineages, and ecologies in epic Darwinian struggles.[1] I have selected a suite of simulations that can be arrayed to correspond to moments in this narrative tendency. The primary exhibits will be Tom Ray's Tierra, at the beginning of the chapter, and John Holland's Echo system, at the end. In the middle of our expedition, I step behind the scenes to examine a programming procedure called the genetic algorithm, a technique that can "evolve" programs in artificial worlds and that serves as scaffolding for many of the systems we will be looking at through the peculiar looking-glass that is the computer screen.[2]

I use the image of the looking-glass because I want to play with a similarity between the way Lewis Carroll's Alice travels through the looking-glass and the way Artificial Life researchers rescue stories from their simulations. Alice, seeing smoke in the mirror, cannot be sure there is fire on the other side until she pretends she can slip herself through the mirror's polished surface. Artificial Life scientists, seeing self-replication, competition, and other phenomena on the surface of their computer screens, cannot be sure there is life on the other side until they cast an apparatus of interpretation across the luminous boundary between the real and the virtual. Inside the computer are the worlds Artificial Life scientists would create and explore. Outside are the social and cultural worlds they inhabit and refer to for tools to make sense of their creations. I intend to show how the inside and the outside mutually constitute one another, taking a kind of Möbius trip through the looking-glass worlds of Artificial Life. Important for this voyage, we must keep in mind that mirrors both reflect and reverse. The artificial life-forms swarming in the artificial worlds of computer simulations both double and invert researchers' images of existing life. Digital creatures stand as *doppelgängers* to the people who program them, as silicon copies both like and unlike their carbon creators.

TIERRA

My circuit of Artificial Life worlds begins with Ray's Tierra. Tierra is very popular in Artificial Life and at SFI (where Ray has maintained an affiliation parallel to his permanent job at the University of Delaware and a recent temporary post at the ATR Human Information Processing Research Laboratory in Kyoto, Japan), and some people, including Ray, think of the "digital organisms" in Tierra as truly alive. Ray has gotten voluminous correspondence about Tierra: people have suggested modifications, asked to use it in biology classes, and requested that he examine codes for digital organisms they have created at home.

   Tierra is a computer program that serves as an environment within which short assembly language programs resident in RAM replicate based on how efficiently they make use of CPU time and memory space.[3] When the program is set in motion, an array of digital organisms emerge, descending in a cascade from an "ancestral" self-replicating program "inoculated" into the system by the user at the beginning. Only the "fittest"—those that replicate most quickly, those that find ways to pirate the replicative subroutines of other programs—survive. According to Ray, Tierra is not simply a model of evolution. It is "an *instantiation* of evolution by natural selection in the computational medium" (1994a:183; emphasis added). Ray's definition of evolution allows him

to enfranchise Tierran organisms into the dominion of life: "I would consider a system to be living if it is self-replicating, and capable of open-
ended evolution" (1992a:372). Indeed, for Ray, the parallels are many
between organic and computer life:

> Organic life is viewed as utilizing energy, mostly derived from the sun,
> to organize matter. By analogy, digital life can be viewed as using CPU
> (central processing unit) time, to organize memory. Organic life evolves
> through natural selection as individuals compete for resources (light, food,
> space, etc.) such that genotypes which leave the most descendants increase
> in frequency. Digital life evolves through the same process, as replicating
> algorithms compete for CPU time and memory space, and organisms
> evolve strategies to exploit one another. (1992a:373–374)

The digital life in Tierra consists of self-replicating assembler language
programs:

> Assembler languages are merely mnemonics for the machine codes that are
> directly executed by the CPU. . . . Machine code is the natural language of
> the machine, and machine instructions are viewed by this author as the
> "atomic units" of computing. It is felt that machine instructions provide
> the most natural basis for an artificial chemistry of creatures designed to
> live in the computer. In the biological analogy, the machine instructions are
> considered to be more like the amino acids than the nucleic acids, because
> they are "chemically active." They actively manipulate bits, bytes, CPU
> registers, and the movements of the instruction pointer. . . . The digital
> creatures . . . are entirely constructed of machine instructions. They are
> considered analogous to creatures of the RNA world, because the same
> structures bear the "genetic" information and carry out the "metabolic"
> activity. . . . The "genome" of the creatures consists of the sequence of ma-
> chine instructions that make up the creature's self-replicating algorithm.
> The prototype creature consists of 80 machine instructions; thus, the size of
> the genome of this creature is 80 instructions, and its "genotype" is the spe-
> cific sequence of those 80 instructions. (1992a:374)

The prototype creature and its descendants exist in a block of memory
called "the soup." Each creature occupies some block of memory and
"reproduces" by copying its code into another block of memory:

> Each creature has exclusive write privileges within its allocated block of
> memory. The "size" of a creature is just the size of its allocated block (e.g.,
> 80 instructions). . . . Each creature may have exclusive write privileges in at
> most two blocks of memory: the one that it is born with which is referred
> to as the "mother cell," and a second block which it may obtain through
> the execution of the MAL (memory allocation) instruction. The second

block, referred to as the "daughter cell," may be used to grow or reproduce into. When Tierran creatures "divide," the mother cell loses write privileges on the space of the daughter cell, but is then free to allocate another block of memory. (1992a:378)

Tierran creatures are nothing more than machine language programs that code for self-replication. The language in which these programs are written was devised for Tierra by Ray, and consists of 5-bit "words" made of 0s and 1s. The words are machine instructions that direct the copying and shifting of program/organism components between registers in the memory. Very roughly, digital organisms copy themselves when they find locations in the memory that are complementary to the patterns of 0s and 1s that make up their own program (if their first few lines read 11001, 11001, 11001, and 11001, for example, they look for templates in memory that read 00110, 00110, 00110, and 00110). They appropriate the templates at these locations (or "addresses") as resources to manufacture a copy of themselves. This matching of complementary patterns is modeled after the way nucleotide bases on strands of DNA pair up. In DNA, adenine pairs with thymine and guanine with cytosine, and it is this pairing that zips strands of DNA together and allows copies to be made from unzipped DNA.[4] The analogy Ray makes between the "genetic language" of DNA and the Tierran "genetic language" is clear:

> The organic genetic language is written with an alphabet consisting of four different nucleotides. Groups of three nucleotides form 64 "words" (codons), which are translated into 20 amino acids by the molecular machinery of the cell. The machine language is written with sequences of two voltages (bits) which we conceptually represent as ones and zeros. The number of bits that form a "word" (machine instruction) varies. (1994a:185–186)

In Tierra, this number is five, so that 32 5-bit combinations of 0 and 1 are possible—32 "words" in the "genetic code" of the Tierran language, all of which mean something. These words are considered analogous to the triplets of base pairs in DNA that code for the amino acids out of which proteins are made. Programs replicate by following the machine instructions of which they are constituted, reading word after word, "codon" after "codon," to first measure their length (finding out where in memory they begin and end), to next set aside that much space in nearby memory to copy themselves, and to finally do that copying and start the whole process over again.

In this world of reading, writing, and reproduction, programs do not reproduce perfectly; if they did, there would be no change, no room for evolution:

> In order to insure [*sic*] that there is genetic change, the operating system randomly flips bits in the soup, and the instructions of the Tierran language are imperfectly executed. Mutations occur in two circumstances. At some background rate, bits are randomly selected from the entire soup (60,000 instructions totaling 300,000 bits) and flipped [from 0 to 1 or vice versa]. This is analogous to mutations caused by cosmic rays, and has the effect of preventing any creature from being immortal, as it will eventually mutate to death. . . . In addition, while copying instructions during the replication of creatures, bits are randomly flipped at some rate in the copies. (1992a:379)

Mutations do not always result in dysfunctional programs, precisely because many different 5-bit patterns of 0s and 1s can mean something. As effective mutations accumulate, a run of Tierra that begins with one ancestral program sees the growth of a population of variants.

This process of replication does not continue indefinitely. If it did, the Tierran soup would soon fill up, leaving no space for replication. So Ray has introduced a program called "the reaper," "which begins 'killing' creatures when the memory fills to some specified level (e.g., 80%). Creatures are killed by deallocating their memory. . . . The effect of the reaper queue is to cause algorithms which are fundamentally flawed to rise to the top of the queue and die" (1992a:379). The reaper steadily eliminates the "oldest" creatures in the soup, but also monitors the population to terminate those programs that have made errors in execution. The whole system operates to produce populations of programs that can efficiently copy themselves, the reaper acting as a sort of natural selection favoring programs that can do this quickly. Ray says that "evolving digital organisms will compete for access to the limited resources of memory space and CPU time, and evolution will generate adaptations for the more agile access to and the more efficient use of these resources" (1994a:184–185). He has claimed that in observations of Tierra, he has seen the emergence of complicated evolutionary dynamics. He has seen the rise of digital organisms that parasitize others, stealing their replicative subroutines: "These ecological interactions are not programmed into the system, but emerge spontaneously as the creatures discover each other and invent their own games" (1992a:393).

Ray holds that Tierra is an alternative biological world and is a satisfying site to examine evolutionary process. Originally trained as a tropical biologist at Harvard in the 1970s (under, among other people, the famous sociobiologist E. O. Wilson), Ray has done extensive work in Costa Rican rain forests but has been frustrated by the pace of evolution there, which happens too slowly to reveal to him the ecological dynamics in which he is interested. Ray sees Tierra as perfect for doing experimental evolutionary biology. One can create perfect records of all events and experimentally tweak initial conditions. Ray has built a tool that records the code (the "genetic sequence" information) of every creature that has existed in Tierra, something he feels gives him an advantage over evolutionists who must work with a spotty fossil record.

Ray argues that getting to know Tierra can help biologists expand their theories about what is necessary and what is contingent in biological systems. There is of course a paradox here, one Ray recognizes (1994a:183) but glosses over. This is that the system must be engineered with basic evolutionary ideas in mind (scarcity of resources, the idea that organisms can only be one place at a time, etc.) but must *also* be general enough to serve as a tool for seeing *new* alternative evolutionary phenomena emerge. One researcher I interviewed said that deciding which lifelike notions to include as basic was "where the 'art' in Artificial Life comes in." The programmer-experimenter must mix intuitions about where science is already correct with programming decisions general enough to allow new phenomena to emerge. Ray describes the process: "We must derive inspiration from observations of organic life, but we must never lose sight of the fact that the new instantiation is not organic and may differ in many fundamental ways" (1994a:183). Ray sees a new "nature" inside the computer, but has engineered this nature to have features with which he is already familiar. It is these features that render the program legible as a simulation of evolution at all. As Hayles writes of Tierra,

> The interpretation of the program through biological analogies is so deeply
> bound up with its logic and structure that it would be virtually impossible
> to understand the program without them. These analogies are not like icing
> on a cake, which you can scrape off and still have the cake. Nor are they
> clothes you can remove and still have the figure. The biological analogies do
> not embellish the story; in an important sense, they constitute it. (1995:421)

It is not simply biological analogies that make this system persuasive for people; ideas about creation, the essence of life, reproduction, and competition also make it work.[5]

It is the Tierran creation tale, the writing of code into virtual life, with which I begin my analysis. Most descriptions of the "evolution" of virtual creatures in Artificial Life simulations skip over the fact that self-reproducing programs must initially be produced by human programmers. I zero in on this moment, mining it for clues to how Artificial Life systems get off the ground.

CREATION STORIES AND MASCULINE MONOGENESIS

I narrate the creation in Tierra as it appeared to me when I used the system myself. While in residence at SFI, I obtained over the Internet a version of Tierra that I ran on the Institute's UNIX workstations. I spent many late nights fastened to the light emanating from my Sun Sparcstation, gazing at tables of statistics and graphs of data generated by Tierra. The visuals that accompany Tierra are unspectacular. There are no little creatures that flit about on the screen; there are only lines of buzzing data from which the user is supposed to construct the histories of generations of self-reproducing computer programs (see fig. 9).

To use the Tierra simulator, one types "tierra" at the user prompt (after compiling the simulator and assembling the initial genome). As soon as one types this word of creation, one is presented with a display like that in figure 9, which includes information about the history of the world as it unfolds. When Tierra is running, these numbers are constantly updating, showing us how many instructions have been executed, how many generations have been cycled through, how many creatures exist in the RAM soup, what the average size of a Tierran organism is (by number of instructions in its program), and so on. Using other utilities, one can see such things as the distribution of different-sized programs in the soup (see fig. 10, a photograph of Tierra as it looked on my customized screen).

The screen text frozen in figure 9 is taken from the very beginning of a run and reports information sampled from the dawn of a Tierran

```
InstExec = 0,005911   Cells =    7   Genotypes = 1     Sizes = 1
Extracted =
InstExeC    =      0   Generations  =    0   Mon May 9  21:08:29 1994
   NumCells =      1   NumGenotypes =    1   NumSizes  =        1
   AvgSize  =     80   NumGenDG     =    1   NumGenRQ  =        1
   RateMut  =   3191   RateMovMut   =  640   RateFlaw  =     9600
```

Figure 9.   Tierra simulator display.

Figure 10.   Tierra on my Santa Fe screen, in XWindows on a Sun
Sparcstation. I am monitoring two separate Tierra runs, one in the upper
window, one in the lower. The upper displays a histogram of different
"genotype" sizes; the lower displays a chart like that reproduced in the text.
The crescent moon and leaves around the edges of the screen are part of my
personalized "environment." Photo by Stefan Helmreich and Chel Beeson.

history. Studying this text, we can see that we are at generation zero
and that there is only one digital organism in the soup, indicated by
NumCells = 1. This is the self-replicating program Ray created from
scratch, the program he used and provides to start up the system. Ray
explains what we are seeing: "Evolutionary runs of the simulator are begun by inoculating the soup of 60,000 instructions with a single individual of the 80 instruction ancestral genotype" (1992a:382).

Ray calls this individual the "ancestor" and describes it as a " 'seed'
self-replicating program" (1992b:37). "Seed" is a common term in computer science, usually used in the phrase "random number seed," refer-

ring to a pseudorandom number used as a starting point for a set of computational processes. Ray's phrase echoes this usage but also evokes the meaning of seed as a germinal entity that has latent within it the potential to develop into a living thing capable of producing more seeds. The choice of the word *seed*—and of the word *ancestor*—is important, for it plants the idea that from the word *tierra* we are witnessing nothing less than life-forms in realized potential.

But the use of *seed* does more than this. The anthropologist Carol Delaney has argued that in cultures influenced by Judeo-Christian narratives of creation and procreation, using the word *seed* to speak of the impetus of creation summons forth gendered images. In the creation tales of these traditions, God, imagined as masculine, sparks the formless matter of earth to life with a kind of divine seed: the Word of creation or *logos spermatikos*. In tales of procreation, males, made in the image of a masculine god, plant their active "seed" in the passive, receptive, yielding, and nutritive "soil" of females, "fertilizing" them (see Delaney 1986, 1991). Creation and procreation in these narratives is "monogenetic," generated from one source, symbolically masculine. "Man" and "God" take after one another. I suggest that the creation in Tierra—and note that *tierra* means "soil" as well as "Earth" in Spanish—symbolically mimics the story of creation in the Bible. The programmer is akin to a masculine god who sets life in motion with a word, a word that plants a seed in a receptive computational matrix, a seed that, in its search for nourishment, organizes an initially undifferentiated "soup."[6] We might see in Tierra images of a symbolically "male programmer mating with a female program to create progeny whose biomorphic diversity surpasses the father's imagination" (Hayles 1994b:125).[7]

Chris Langton has claimed that Artificial Life is "the attempt to abstract the logical form of life in different material forms" (quoted in Kelly 1991:1), a definition that holds that formal and material properties can be usefully partitioned and that what matters is form. But form and material, like seed and soil, also have gendered valences for those of us swimming down the stream of Western natural philosophy and life sciences. Aristotle proclaimed in his *Generation of Animals* that in procreation, "the male provides the 'form' and the 'principle of movement', the female provides the body, in other words, the material" (I.XX.729a). Judith Butler notes that "the classical association of femininity with materiality can be traced to a set of etymologies which link matter with *mater* and *matrix* (or the womb)" (1993:31). This gendering of form and material works hand in glove with the seed and soil metaphor.[8] And images of "form" and "seed" easily overlap in Artificial Life when practitioners make analogies between computer code (information) and genetic code.[9] When Tom Ray writes of single-handedly creating digital life in Tierra with a seed, when he remarks that this "digital life exists in a logical, not material, informational universe" (1994a:183), and when he asserts that he occupies the position of god with respect to Tierra, it is hard not to hear echoes of a masculine monogenetic creation.

When one attends closely to the language in Tierra, however, one notices that Ray refers to the ancestor as a "mother" and to descendants as "daughters." This would seem to add a twist, but reproduction remains about transmission of form, though this is now democratized to females. Parentage has become defined as an informational relation, just as "real" parents in the age of in vitro fertilization and surrogacy have become legally synonymous with "genetic" parents, a move that ensures that mothers and fathers are "equal" on terms set by a masculine model (see Paxson 1992; Hartouni 1997). And for all the "femininity" of the Tierran "seeds," they still engage in "begetting": Tierra allows users to access a tool called the genebanker, which in Ray's words at one conference "keeps track of who begat whom."[10] "Begetting" is a word used in chapter 5 of the Book of Genesis to chronicle the passage of "seed" down generations of men. *Webster's Third New International Dictionary* defines *beget* as "to procreate as the father: SIRE."[11]

Ray is hardly the only one to use images of "seeds." The programmer Jeffrey Putnam writes that his simulator, Stew, sparks up when "the world is created and seeded" (1993:948). In a talk at SFI, Mark Bedau opened a report on an ecological simulation by saying, "This system is seeded with a population of handcrafted ancestors." Chris Langton has said of a simple CA instantiation of a self-reproducing automaton, "These embedded self-reproducing loops are the result of the recursive application of a rule to a seed structure" (1989:30). The manual for SimLife instructs the user in "spreading your seed" (Bremer 1992:51). A computer-animated short called "Panspermia" included in the Artificial Life II video proceedings makes blisteringly clear the masculine imagery guiding the seed concept (see Sims 1992). In this film, we see an enormous seed fly through outer space, crash on a barren planet, and explode like a piñata into a plethora of botanical life-forms that ultimately give rise to phallic plants that shoot more seeds into space in cannonlike explosions that suggest energetic ejaculations. In many simulations, programmers make the masculinity of the seed explicit when they name ancestral programs "Adam." The original "seed structure" in Chris Langton's loop is called "Adam" (Farmer and Belin 1992:824), as is an ancestral agent referred to as a "seed" in David Ackley and Michael Littman's "AL" artificial world (1992a:498, 1992b). In a speculative discussion of delivering a package of self-reproducing mineral extraction factories to the moon, Richard Laing tells us of a " 'seed' perhaps weighing one hundred metric tons delivered to an extraterrestrial planetary surface" (1989:58). He reminds us of the masculine principle behind this robot seed with his mention of an automaton offspring receiving "its patrimony" (1989:54).

The possibility that the Tierran experimenter has some resemblance to a god who monogenetically creates life out of nothing is suggested in the annotated assembly code file for the ancestor. This file, containing the "genetic code" of the ancestor, is reproduced in figure 11. The file comes from a directory called the genebank, a sort of digital cryonics lab that stores all genomes evolved in the system for later resurrection in experiments.[12] On the second line of the ancestor's genome file is a field that reads "parent genotype." In most digital organisms, this indicates the program's immediate ancestor. There is no ancestral program for the ancestor, since it was created by Tom Ray. So where most creatures have a parent genotype like "0045aab," the ancestor has the rather peculiar "0666god," a designation that suggests that the programmer is a kind of Faustian figure, playing at a devilishly digital divinity. Because Tierra cannot be understood without knowing that it was intentionally written by a human creator, both evolutionary language and theological language have become necessary to make sense of the system. Ray is clearly being whimsical here, but his jokes perform crucial work in constructing the programs in Tierra as created life.

Themes of masculine creation are not idiosyncratic to Tierra; they permeate the public image of Artificial Life. The popular press often calls Chris Langton "the father of Artificial Life," a phrase that sounds innocent enough—he is merely a man founding a new discipline—until one realizes that we rarely hear "mother" used in a similar way and that the phrase "the mother of artificial life" would sound odd, most readily calling up images of women birthing babies grown from zygotes made in laboratories.[13] Langton has written that Artificial Life is "life made by *man* rather than by nature" (Langton 1989:2; emphasis added), a phrase that depends for its efficacy on a noun for humanity that is not gender

```
format: 3  bits: 2156009671  EXsh     TCsh     TPs      MFs      MTd      MBh
genotype: 0080aaa  parent genotype: 0666god
1st_daughter:  flags: 0  inst: 827  mov_daught: 80              breed_true: 1
2nd_daughter:  flags: 0  inst: 809  mov_daught: 80              breed_true: 1
Origin: InstExe: 0,0  clock: 0  Wed Dec 31 17:00:00 1969
MaxPropPop: 0.8306  MaxPropInst: 0.4239 mpp_time: 0,0
ploidy: 1  track: 0


track 0: prot
          xwr
nop1    ; 010 110 01   0          nop1    ; 010 110 01   56
nop1    ; 010 110 01   1          nop0    ; 010 110 00   57
nop1    ; 010 110 01   2          nop0    ; 010 110 00   58
nop1    ; 010 110 01   3          inc_a   ; 010 110 08   59
zero    ; 010 110 04   4          inc_b   ; 010 110 09   60
not0    ; 010 110 02   5          jmp     ; 010 110 14   61
shl     ; 010 110 03   6          nop0    ; 010 100 00   62
shl     ; 010 110 03   7          nop1    ; 010 100 01   63
movcd   ; 010 110 18   8          nop0    ; 010 100 00   64
adrb    ; 010 110 1c   9          nop1    ; 010 100 01   65
nop0    ; 010 100 00  10          ifz     ; 010 000 05   66
nop0    ; 010 100 00  11          nop1    ; 010 110 01   67
nop0    ; 010 100 00  12          nop0    ; 010 110 00   68
nop0    ; 010 100 00  13          nop1    ; 010 110 01   69
sub_ac  ; 010 110 07  14          nop1    ; 010 110 01   70
movab   ; 010 110 19  15          popcx   ; 010 110 12   71
adrf    ; 010 110 1d  16          popbx   ; 010 110 11   72
nop0    ; 010 100 00  17          popax   ; 010 110 10   73
nop0    ; 010 100 00  18          ret     ; 010 110 17   74
nop0    ; 010 100 00  19          nop1    ; 010 100 01   75
nop1    ; 010 100 01  20          nop1    ; 010 100 01   76
inc_a   ; 010 110 08  21          nop1    ; 010 100 01   77
sub_ab  ; 010 110 06  22          nop0    ; 010 100 00   78
nop1    ; 010 110 01  23          ifz     ; 010 000 05   79
nop1    ; 010 110 01  24
nop0    ; 010 110 00  25
nop1    ; 010 110 01  26
mal     ; 010 110 1e  27
call    ; 010 110 16  28
nop0    ; 010 100 00  29
nop0    ; 010 100 00  30
nop1    ; 010 100 01  31
nop1    ; 010 100 01  32
divide  ; 010 110 1f  33
jmp     ; 010 110 14  34
nop0    ; 010 100 00  35
nop0    ; 010 100 00  36
nop1    ; 010 100 01  37
nop0    ; 010 100 00  38
ifz     ; 010 000 05  39
nop1    ; 010 110 01  40
nop1    ; 010 110 01  41
nop0    ; 010 110 00  42
nop0    ; 010 110 00  43
pushax  ; 010 110 0c  44
pushbx  ; 010 110 0d  45
pushcx  ; 010 110 0e  46
nop1    ; 010 110 01  47
nop0    ; 010 110 00  48
nop1    ; 010 110 01  49
nop0    ; 010 110 00  50
movii   ; 010 110 1a  51
dec_c   ; 010 110 0a  52
ifz     ; 010 110 05  53
jmp     ; 010 110 14  54
nop0    ; 010 110 00  55
```

Figure 11. File containing the genetic code for the Tierran "ancestor" program.

Figure 12.   Poster for the second Artificial Life conference. Copyright 1989
Chris Shaw. Reproduced with permission.

neutral ("life made by people rather than by nature" would not work
equally well). The notion that Man replaces God and renders Woman
irrelevant in the new creations of Artificial Life is vividly illustrated in
figure 12, a reproduction of the poster for the second conference on Artificial Life, in which a white male programmer touches his finger to a
keyboard to meet the waiting fingers of a skeletal circuit-based artificial
creature. The programmer is levitating, symbolizing his transcendent position. This Escheresque quotation of Michelangelo's Sistine Chapel rendering of the creation of man, in which God touches the extended index
digit of the first man, Adam, condenses many of the themes of creation
I have examined here.

Imagery of a masculine creation also surfaces in researchers' casual comments, jokes, and, occasionally, confessions about why they do Artificial Life at all. The links between masculinity, paternity, and the creation of Artificial Life worlds were evoked for me one day when, at an Institute workshop, a male researcher claimed to have a "grandfatherly pride" in a program he had had the inspiration for but had not himself programmed. The symbolically masculine creation of silicon life is a theme some men in Artificial Life explicitly play with; some joke that their wives take care of the kids while they take care of the virtual creatures. Craig Reynolds, in acknowledgments for an article in *Artificial Life IV,* writes, "Special thanks to my wife Lisa and to our first child Eric, who was born at just about the same time as individual 15653 of run C" (1994:68). Ray quotes his wife, Isabel Ray, in one epigraph: "I'm glad they're not real, because if they were, I would have to feed them and they would be all over the house" (Ray 1994a:202).[14] Steven Levy's pop description of Ray's "creation of life" reruns a Frankensteinian tale of male creation: "On January 3 [1990], working at night on a table in the bedroom of his apartment while his wife slept, Ray 'inoculated' the soup with his single test organism, eighty instructions long. He called it the 'Ancestor'" (1992b:221).

Normative notions of fathering in Euro-American culture also inform the ways Artificial Life researchers play God. Artificial Life researchers are creators who can inseminate virtual worlds and leave. Creatures need no feeding or nurturing, and they are supposed to grow up quickly and become independent. *WIRED* editor Kevin Kelly, in an interview in *Microtimes,* directly compares creating Artificial Life to bringing up kids. Even when he speaks of nurturing, he uses the language of programming:

> Anybody with kids knows that kids can be monsters. . . . So how do you keep it from being a monster? The answer is not that you don't allow it to be born. The answer is not that you totally control it. The answer is that you train it up, you manage it, you instill certain operating principles, you guide it—and then you let it go. I think that it's the same thing with technology. . . . [Y]ou would *love* it to surprise you; you would love a child to come up with something new or original. (Quoted in Walsh 1994:328)

In spite of the fathering motifs quilted into Artificial Life talk and programming, some people I interviewed wondered whether Artificial Life might be seen as an expression of male researchers' birth envy. Helena was one:

> Women create things, right? We have babies and we certainly know the role of males in that, but it's not clear how much men feel that role, and

maybe that's what ALife is. Maybe men would like to give birth to something and here it is, this is it. They're saying to us, "We're going to beat you guys. We're going to create entire worlds."

Andrew, a computer scientist, did in fact tell me he created artificial worlds in part because he felt frustrated he was not a woman and could not create "naturally" (by birthing). Another person suggested that if pressed to account for the fact that there were more men in Artificial Life than women, he would "propose the theory that men are more frustrated in the urge to create life than women, and that ALife gives an outlet to this frustration." At one conference, I met a young man who had a remarkable set of reflections on this topic:

> In the middle ages, male alchemists tried to come up with ways to bypass women in reproduction. I was thinking that Artificial Life research could very easily be just another way of being a surrogate, for males to bear children knowing that they actually can't. I'm not sure whether I believe it, but it has occurred to me. It reminds me of something which no one has yet asked me, but which I have thought about—and I still haven't come up with an answer—which is: why exactly it is that I'm interested in Artificial Life and how I can reconcile that with the fact that I'm gay. Of course, the mistake that many people make is that they assume that anyone who is other than straight is going to incorporate their sexuality into everything they do, which isn't necessarily the case. I am interested in the idea of evolution and reproduction. I've never particularly been interested in *sexual* reproduction. I don't know if that's an artifact of my sexuality or not. Who knows? I've been asked questions having to do with the evolution of sexuality, and I've thought about whether it relates to my interest here. I'm not sure that it does. I'd really be hard-pressed to say that I have any opinions about whether my interest in Artificial Life is because I'm not actually going to reproduce. I don't think that's it really. It would be very difficult to find that kind of link. It's amusing to think that Artificial Life is overrun by males because it's their way of having babies. It's a fun sort of idea, but I certainly don't have any vested interests in proving that it's true.

This man's ironic and reflexive reflections reveal an intriguing inconsistency. Artificial Life is figured as a practice in potential dissonance with (normatively nonreproductive) gay masculinity but is simultaneously construed as something in which men in general—and perhaps gay men in particular—should be interested.

In all these pronouncements, male creation is imagined as fundamentally artificial and female creation as fundamentally natural. Men create artificial life; women create natural life. There is a curious contradiction here. On one side, females create "naturally," and birthing is conflated with reproduction, with males vanishing from the scene. On the other side, males are the sole creative force in creation and procreation, with feminine contributions figured as simply supportive. Female birthing is everything at one moment and nothing in the next[15]—so much nothing that reproduction can proceed without women, can even be pristinely transferred to a different vessel, the computer.[16] Some Christians believe that the pure and uncorrupted virgin Mary was the perfect vessel for the seed of God, birthing a child who was not half-God, half-Mary, but all God. Computers might be seen as capable of the same clean conception as Mary, bearing faithfully those formal self-reproducing seed programs that are the conceptions of Artificial Life scientists.[17]

Stories of masculine creation usurping or bettering female creation can be found in many scientific narratives. The physicist Brian Easlea (1983) has written of how male nuclear weapons scientists often speak of the bombs they produce as babies and has interpreted this as bespeaking the desires of a masculine science to appropriate and transcend female reproductive abilities. The anthropologist Hugh Gusterson (1996), in his ethnographic study of weapons scientists, has argued that while there is something notable in how male researchers use this language, it is ultimately unconvincing as a key to their inner psychology. After all, women can easily use this language, and sometimes do (though I found this rare among people I spoke with). The language does not reflect subconscious motives so much as it allows work to get done, so much as it draws on shared imagery to provide a lexicon for producing artifacts. This is not to say the language is strictly utilitarian; it also reproduces problematic gender ideologies and structures of feeling and reasoning about relations between women and men. In an effective way, masculine God imagery allows researchers to imagine that they are creating life in fertile but empty worlds. Of course, if this imagery were all there is to it, Artificial Life could not function without it. There are a number of other images that enable computer programs to be described as alive. The next one I examine manufactures programs and organisms as synonyms.

ORGANISM AS INFORMATION

What sustains Tierra as a serious piece of scientific work is a view that programs are good models for organisms, such good models that they can be considered candidates for *being* organisms. What makes this sensible? An anatomy of the Tierran organism begins to answer the question. Returning to the genome of the ancestral Tierran creature, we see that the digital organism is a string of information, a sequence of os and 1s interpretable as a set of program instructions. Ray writes, "The 'body' of a digital organism is the information pattern in memory that constitutes its machine language program" (1994a:184).

Many Artificial Life formulations imagine life process as program. In their discussion of RAM, a simulation platform that has been used to simulate lek formation in sage grouse and predator-prey relations between foxes and rabbits, Taylor et al. write, "RAM is based on the observation that the life of an organism is in many ways similar to the execution of a program, and that the global (emergent) behavior of a population of interacting organisms is best emulated by the behavior of a corresponding population of coexecuting programs" (1989:275; see also Hogeweg 1989). One researcher summarized to me the informatic view of organisms: "After a while the analogy between self-replicating programs and living organisms becomes so perfect that it becomes perverse to call it merely an analogy. It becomes simpler just to redefine the word *organism* to apply to both chemical and software creatures. What they have in common is much more important than how they differ." Andrew was so persuaded that organisms were computations that he remarked to me, "My researches have been stabs at the space of all possible computations, of which I remain convinced that I am one. I can't see what else I could be."

This identification of organisms with programs does not merely indicate a commitment to form over matter, it plays on the popular and scientific conceit that organisms are ultimately nothing other than the unfolding of a genetic plan and that genes equal information structures analogous to computer programs. The idea that organisms can be collapsed into genes motivates Ray's contention that "the bit pattern that makes up the program is the body of the organism and at the same time its complete genetic material" (1994a:185). Ray argues that his organisms are analogous to hypothesized early life-forms made entirely of RNA; this may be so, but there are cultural histories and beliefs that make possible a preference for identifying organisms with genes and genes with information.

The idea that genes make the organism is a philosophy known as genetic determinism. Genetic determinism has its origins in beliefs that there are essential, inborn differences between organisms and that these can be traced to hereditary material. There are roots for genetic determinism even in pre-Mendelian biology; in the seventeenth and eighteenth centuries, spermist preformationists held that each human and animal sperm contained a miniature creature that simply grew in size as it was nourished by an egg (see Tuana 1989). In the early twentieth century, the discovery that mammalian gametes (sperm and eggs) were "sequestered" from the rest of the body—that is, were not subject to change in an organism's lifetime except through mutations—led some to posit that differences between groups must be grounded in heredity. All species of genetic determinism embody metaphysical commitments to essential hereditary difference, and, not surprisingly, advocates of genetic determinist positions often seek to rationalize or "explain" social difference and inequality (see Kevles 1985). Fetishes of the hereditary material have shaped an intense focus on genes in evolutionary and population biology. The various genome projects mark an apex of this kind of thinking.

In mid-twentieth-century molecular biology, scientists likened the hereditary material to a code. Modern molecular genetics developed in dialogue with computer science; many of the people who used computers during World War II (for physics problems, decryption projects) later turned their attention to problems in molecular biology. The physicist Erwin Schrödinger's contention in his influential *What Is Life?* that "chromosomes . . . contain in some kind of code-script the entire pattern of the individual's future development and of its functioning in the mature state" was a key image in the development of molecular biology (1944:20). It became commonplace to speak of a "genetic code," and problems of embryogenesis and development took a back seat to "cracking" the genetic code, commonly understood to contain the "secret" of life itself (see Keller 1992b, 1995). An organismic biologist who was annoyed with this state of affairs and bothered by the identification of programs with organisms reflected, "In modern biology, organisms have disappeared as fundamental entities; they've been replaced by genes and their products, which is a logical consequence of Darwin's focus on inheritance and natural selection as the primary features of evolving systems. And molecular biology in the twentieth century has just followed this to its logical conclusion and the result is you have genes as the essence of life."

Genetic determinism and images of DNA as code have made possible the fashioning of programs as life-forms. But they also continue more profound histories. The philosopher Susan Oyama locates the idea that genes are a "blueprint" for an organism in a Western metaphysical tradition of separating form from matter, of assuming that ontogeny is the playing out of a developmental "program" and that "information . . . exists before the interactions in which it appears" (1985:27).[18] Doyle (1997b) writes that under the spell of understanding DNA as a code-script, twentieth-century biology has conflated vitality and textuality, and that Artificial Life is a recent symptom of this. The implosion of life and information has rendered it possible to see programs as dead ringers for living things.[19]

The idea that genes are programs for organisms has been adopted by many evolutionary biologists. The prominent evolutionist Ernst Mayr summarized the notion nearly twenty-five years ago: "The young in some species appear to be born with a genetic program containing an almost complete set of ready-made, predictable responses to the stimuli of the environment" (1976:23). Mayr is a formidable figure in biology, and Taylor et al. cite this quotation favorably in their RAM article (1989). Perhaps the most striking identification of the organism with its genes, and of genes with replicating information structures, was propagated by the evolutionary biologist Richard Dawkins in *The Selfish Gene,* in which he wrote, "We are the survival machines—robot vehicles blindly programmed to preserve the selfish molecules known as genes" (1976:ix). Dawkins's phrasing of the issue itself depends on a computational image, illustrating the self-referential character of an Artificial Life that borrows ideas from a biology already saturated with computational analogies. Dawkins has been an exponent of a brand of evolutionary biology known as sociobiology, a strand of theory first made popular in the 1970s. Sociobiology is a kind of neo-Darwinism, a modern phrasing of Darwinian theory in the language of population genetics, and it takes seriously a view of evolution as a process in which genes act to maximize their fitnesses. Along with E. O. Wilson, Dawkins is the most widely read advocate of this highly formal and mathematized view. Dawkins's idea that evolution is equal to the differential reproduction of informatic entities was mentioned repeatedly by people I interviewed, and his books were a feature of many scientists' libraries. Not surprisingly, Dawkins has become interested in Artificial Life. He has written a program called Blind Watchmaker, in which the user acts as an agent of artificial selection on "biomorphs," simple geometric computer shapes coded for by elementary programs (see Dawkins 1986, 1989). He has published in

Artificial Life proceedings, and continues to take an active role in promoting the field.

The vivification of computer programs is also enabled by the infiltration of genetic definitions of life into everyday talk. Artificial Life researchers think through these notions not just in science but in other domains as well. Consider the ways researchers responded to questions about abortion and the debate in the United States about "when life begins." All of the people I spoke with felt strongly that women had a political right to choose whether to follow a pregnancy to term, but all also believed that "life" began at conception. One researcher said, "I think abortion is really a horrible thing. I think life has started there, and it's analogous to squashing a little tiny seedling. Life has started the moment after conception." Another said to me that a zygote is alive because of its genetic potential, a potential he understood as a program: "You start out with a fertilized egg, and yes, it's alive. It is definitely alive. It's a living cell, and it's programmed to develop into this fabulous complicated thing you call a baby. So there's no doubt that this is alive." It is this view of life as genetic potential that allows people to say that "life" begins at conception, the genetic fusion of the gametes (see Franklin 1993b). When Tom Ray pleads for an Internet-wide version of Tierra by saying that we've got to give life a chance, he, like people in the "pro-life" movement, identifies "life" with genetics.

This identification of organisms with genes is powerful enough that people use it to reflect on their own lives. Many have taken to heart Dawkins's notion that humans are automata designed to pass on genes. At an SFI workshop on computation and evolutionary biology, one man remarked, in a room full of men only, "Here we are sitting around the table talking, when all we're about is replicating." This statement points to a strong identification with one's genes, one that discourages identification with the embodied self as the essential self, and one that entirely forgets about pregnancy—just like definitions of "life" that say it begins at the moment of conception. The organism is flattened into its genes and loses the body in the bargain. How does this come to happen? I suggest that it is wrapped up in part with masculinist denials of things bodily.

A few men I interviewed told me they thought Artificial Life may usher in a stage of evolution in which carbon bodies will be rendered obsolete by the super-durable bodies of extraterrestrial robots or the virtual bodies of cyberspace critters, bodies into which some claimed we humans may well download our consciousness.[20] Here, the mind, like the gene,

is a program, an information structure and process capable of reproducing itself through appropriate robot surrogates.[21] Some of the most
incredible stories of disembodiment have come from Hans Moravec, a
roboticist who has published in Artificial Life proceedings and is famously the author of *Mind Children* (1988). Moravec has been recorded
as wishing he were a robot, free of the limits of earthly existence. Helena reported that this view frightened her: "When I see someone like
Moravec, who says that our progeny will be these machines, we're talking about something scary, which is a turning away from emotional reality. That means a lack of contact with it, so that you don't know what
it is. At the same time, that lack of knowledge about it is inducing strong
emotional reactions like fear, fear of the social and emotional." In an exchange with Steven Levy, Danny Hillis, a computer scientist, expresses
a view of the value of human life as rooted not in bodies but in minds:

HILLIS: . . . We're a symbiotic relationship between two essentially differ-
           ent kinds of things. We're the metabolic thing, which is the mon-
           key that walks around, and we're the intelligent thing, which is a
           set of ideas and culture. And those two things have coevolved to-
           gether, because they helped each other. But they're fundamentally
           different things. What's valuable about us, what's good about hu-
           mans, is the idea thing. It's not the animal thing.

LEVY: Right. If some alien being took a human egg and gestated it some-
           where away from any human contact, what you got wouldn't be
           what you'd want to call human.

HILLIS: Exactly. So to me, if we can improve the basic machinery of our
           metabolism [the result can still be "human"]. See, I think it's a
           totally bum deal that we only get to live 100 years. I think that's
           awful, that's barely enough chance to sort of get going.

LEVY: How old are you?

HILLIS: Thirty-three. And I want to live for 10,000 years. . . . If I can go
           into a new body and last for 10,000 years I would do it in an in-
           stant, no second thoughts.

                                                          Levy 1992a:39

There are historical reasons why this imagery is easily adopted by men
in positions of social privilege. In Cartesianist science, mind and body
have been separated, and while the mind has been symbolically associated with the formal, the essential, the rational, the scientific, and the
masculine, the body has been accociated with the material, the supereroga-
tory, the superstitious, the irrational, and the feminine. Scientific "objectivity" has consisted in describing "formal" or structural rather than
"material"phenomenological qualities. This epistemological prejudice has

been supported culturally by a notion that men are less leashed to their bodies than women and can therefore be more "objective" (see Keller 1985; Harding 1991). As the philosopher Sandra Harding has argued, men have often been able to deny their embodiment, or regard it as ontologically trivial, because so many of their bodily needs have been taken care of by the (often invisible) labor of women. This has been a general feature of Western society, and most people have encountered it in office, family, household, and school situations. At SFI, a not untypical site of normative office politics and scientific work, researchers' corporeal and daily needs and wants are taken care of to such an extent that they may forget that it takes real physical work to satisfy them. Food is catered, faxes are sent, mail is sorted and distributed, conferences are coordinated, toilet paper is magically replenished, and offices are cleaned. And gender matters here. Although there are certainly exceptions, many of the people doing the work of Artificial Life simulation at SFI are men, and most staff supporting the bodily and worldly needs of the researchers are women. Many men who work at the Institute have gotten used to being cared for and cleaned up after in other realms of their everyday lives and have put low priority on domestic tasks (something testified to by email warnings from female staff to researchers, most of whom are male, about keeping the SFI kitchen clean). The structures of social inequality that have made this the case have been rationalized by and continue to support a science in which males are taken as the paradigmatic knowers, ideally unencumbered by and unconcerned with the messiness of embodied life. Following Harding (1991), who asked what the laws of nature would look like if they were discovered by those who had to clean up after them, I ask what people's theories of the essence of life would look like if they spent their time taking care of the everyday needs of organic systems. I would wager that people might see just how far it is between a genetic code and a full-bodied organism.

But Artificial Life forms are clean of the untidiness of embodiment. Ray's Tierra may halt because the reaper is not killing digital organisms quickly enough, but not because the soup is becoming choked with blood or excrement. With Artificial Life, biology will be as pristine as mathematics. Chris Langton writes, "Computers should be thought of as an important laboratory tool for the study of life, substituting for the array of incubators, culture dishes, microscopes, electrophoretic gels, pipettes, centrifuges and other assorted wet-lab paraphernalia, one simple-to-master piece of experimental equipment devoted exclusively to the in-

cubation of information structures" (1989:39). The only way this statement makes sense is if we see life as information and understand bodies
simply as messy instantiations. In Langton's tableau, all the things difficult to "master" are associated with wetness, a quality symbolically
linked to the feminine (Theweleit 1977) and to the confounding of
boundaries (Douglas 1966).

The idea that self-replicating information structures might be alive
finds intriguing expression in debates about "computer viruses," which
are named on analogy to biological viruses. A virus is defined as "a segment of machine code (typically 200–4000 bytes) that will copy itself (or
a modified version of itself) into one or more larger 'host' programs when
it is activated. When these infected programs are run, the viral code is
executed, and the virus spreads further" (Spafford 1994:250). Many Artificial Life researchers point to computer viruses as evidence of the spontaneous emergence of artificial life. Here, they say, are entities that fulfill the formal qualities of viral entities, and here are entities detected and
named some four years before the formation of Artificial Life.[22] Biologists disagree about whether viruses are alive, and Artificial Life scientists have appropriated this uncertainty to argue about whether computer viruses might not be borderline cases of artificial life. Figuring
computer viruses this way allows researchers to claim that there is in fact
a threshold to be crossed.[23] And this brings us to a discussion of how researchers construct their creations as "primitive."

PRIMITIVITY IN THE NATIVITY
OF ARTIFICIAL LIFE

Artificial Life researchers frequently describe artificial organisms as
"primitive" or "in their infancy." In a way, this is very sensible. Artificial Life scientists are just getting started. One could hardly expect them
to instantly create artificial life equivalent to the most complicated living things we know. But such modest claims are also part of a rhetorical strategy that enables researchers to argue that they are actually on
the path to creating full-fledged life. We should be curious about the images Artificial Life researchers use to render their creatures primitive, for
they often refer us to religioscientific narratives of origin and to images
of the primitive as other to the white man.

Artificial worlds are often outfitted as Gardens of Eden. The phrase
"Garden of Eden" is even used to refer to one kind of initial CA state

(see Wuensche 1994). The computer scientist David Jefferson and his co-workers at UCLA name one simulation program Genesys, an explicit reference to the biblical tale of creation (see Jefferson et al. 1992). In some simulations, there is an exile from the garden, as initially bountiful resources become scarce and the artificial world transforms into a hostile place. In a video on Ackley and Littman's AL system, Ackley interprets a run of the simulator in a way that suggests a paradise lost: "As the thousands of time steps pass by, the agent population begins to grow—and with such success come new problems. The world's not as roomy as it used to be. There's more crowding, there's more competition" (Ackley and Littman 1992b).

Parallel to and in hybridity with biblical imagery of early life are scientific tropes of origin. Thus Artificial Life workers speak frequently of "primordial soups." Larry Yaeger writes that "PolyWorld may be thought of as a sort of electronic primordial soup experiment (1994:264; see also Putnam 1993). The computer scientist John Koza, in introducing a technique called "genetic programming," writes, "This chapter addresses the question as to whether self-replicating computer programs can spontaneously emerge from a primordial ooze of primitive computational elements" (1994:226). He continues, "Spontaneous emergence of computational structures capable of complex behavior might occur in a sea of randomly created computer program fragments that interact in a random way. Such a sea and the turbulent intermixing of entities in this sea of entities has some degree of biological plausibility since biological macromolecules randomly move and randomly come into contact in their cellular milieu" (1994:236). These images rely on images of life coming from disorganized soups, oozes, and seas (we might add the "soil" of Tierra to this list of disordered, feminized forms of primal "nature"). In Ray's writings, we learn that we should think of digital organisms in Tierra as akin to "hypothetical, and now extinct, RNA organism[s]," which were "presumably nothing more than RNA molecules capable of catalyzing their own replication" (Ray 1994a:185).

While early life-forms are popular in Artificial Life modeling, researchers also craft computer creatures in the mold of existing life, especially that life they consider elementary. Hence there are numerous simulations with populations modeled after ants (e.g., Collins and Jefferson 1992; and see Koza 1992), bees (Hogeweg 1989), or fish (Terzopoulos, Tu, and Grzeszczuk 1994; Terzopoulos, Rabie, and Grzeszczuk 1997). Rodney Brooks and his students at MIT's Mobile Robot Lab create robot knockoffs of insects. Some researchers name their programs after animals even if they have only the barest resemblance to the species in question. Simple programs represented as trapezoids wired with circuits are called "turtles" (Travers 1989), and triangles that aggregate into patterns resembling "flocks" go under the pun name "boids" (Reynolds 1992). All these models are surrounded with caveats about their status as caricatures of real life. We are often told that such models only capture aspects of the behavior of the animals by which they are inspired. But the use of insect and animal images suggests that programs are at the same level of complexity as simple organisms from the real world— or soon will be. The privileged status of simple animals in Artificial Life creates a sense that researchers are starting from basics, and does so because of an implicit assumption that simple creatures of today are analogous to the ancestors of contemporary complex creatures. This assumption is suspect, as many evolutionary biologists have pointed out, for it preserves the concept of a great chain of being with humans at the top (see Dawkins 1992).

Larry Yaeger makes it clear that modest artificial creatures are promises of bigger and better things:

> While one of the grand goals of science is certainly the development of a functioning human level (or greater) intelligence in the computer, it would be an only slightly less grand achievement to evolve a computational *Aplysia* that was fully knowable—fully instrumentable and, ultimately, fully understandable. And perhaps it is only through such an evolutionary approach that it will be possible to provide the important milestones and benchmarks—sea slug, rat, simian, . . . —that will let us know we are on the right scientific path toward that grander goal. (1994:294)

The grand project of Artificial Intelligence was to create artifactual minds equal to or better than human minds. This project was the apotheosis of Cartesianist thinking. If the project of Artificial Intelligence were successful, the body could be permanently split from the mind. The mind would finally be liberated from the sticky, limited, and overly emotional flesh. One could easily argue that Artificial Intelligence was conceived in the image of the rational white European man, ideally a calculating, objective, and reasonable entity, constructed in large part by naming others as subjective and irrational. A stated project of Artificial Life is to produce Artificial Intelligence by bringing elementary bodies back in (even if, as we have seen, they are often steamrollered into their genes), subjecting them to virtual evolutionary forces that will press them to

develop intelligence from first principles. They will learn to evade predators before becoming chess masters. Owing to this notion that Artificial
Life organisms must be more primitive than contemporary organic life,
descriptions of such organisms sometimes reflect what Artificial Life
workers—who think of themselves as quite advanced entities—think
they are not.

The anthropologist Michael Taussig (1993) argues that in the history
of Western attempts to construct automata that mimic rudimentary behaviors associated with life, people have made machines in the image of
that which they have considered simpler than themselves. Since the creators of automata have almost always been fully grown white men existing in a world in which women, children, dark-skinned people, people from the "East," and animals were marked as primitive, such
automata have been cast as facsimiles of white children or women playing music, "negro minstrels" strumming guitars, "Turkish" people playing chess, and ducks drinking water. Present-day Artificial Life carries
this legacy forward. In reports on the simulated universes of Artificial
Life, we read epic histories of organisms that must remind us of a sort
of *National Geographic* history of humanity. In Ackley and Littman's
"crowded" AL system discussed above, organisms start to "eat" each
other when they get too cramped. Ackley's video narration tells us that
"when it gets crowded down there against the wall, the agents start running into each other. They're killing and eating each other down there,
offspring eating parent, and they're learning to like it." Ackley proposes
that this set of agents is participating in a "cannibalism cult." After the
cult eats itself to death, later organisms exhibit behavior rational enough
that Ackley and Littman propose that "the potential exists in AL for sophisticated agents with shielded plant-learning genes to discover agri-
culture" (1992a:506–507). In a situation that arose in his PolyWorld
simulation, Yaeger happened on a "species" of organisms that never
moved and that "ate" others of its kind. He referred to this variety as
"indolent cannibals" (1994:283), noting that "these organisms . . . kill
each other, and eat each other when they die" (1994:283). Such images,
though offered with a knowing mischief, are highly racialized, since
"cannibals" have often been associated with images of black Africa and
indigenous America.

The primitivity of Artificial Life organisms is also constructed through
a sense that they are harmless. People often refer to simulation creatures
as "little," or as "these little guys." And sometimes, as in Tierra, they

describe them as female. In the history of ideas about artificial life, creatures created male frequently rise up against their male creators (as in *Frankenstein*), while creatures created female (as in the Pygmalion tale) often fall in love with their male creators. Descriptions of artificial life creatures as female, coupled with descriptions of them as dutifully reproductive, construct in a patriarchal imagination the idea that these creations are benign and perhaps even "lovable" or, as one programmer phrased it to me, possessed of "an endearing quality." Andrew's statement that his creatures were "wimpy, pitiful little life" indicates that they are simple by virtue of being weak and impotent, perhaps feminized.[24] This image of what counts as artificial life worth making was summarized by a person who told me, in a comic riff on the master/slave dialectic circulating in much Artificial Life, "As a lazy hedonistic bachelor, I want domestic machines which will work constantly without my intervention to make my place palatable. In case company drops by."

Although no one strives to produce creatures that are "wimpy," many researchers at least hope that the creatures they make will be "cute." Artificial Life researchers and fans experience artificial organisms as cute most viscerally when presented with elaborate simulation portraits of artificial organisms approximating well-known lifelike behaviors. If graphics are good enough, and if programmers emphasize that behaviors are simulated and not scripted, figures on screens can look as though they are trying very hard to mimic well-known behaviors (indeed, one meaning of *cute*, according to *Webster's Third New International Dictionary,* is "obviously straining for effect"). At one Artificial Life conference, the computer scientist Karl Sims gave a talk in which he showed a video of simulated creatures with boxes for arms, legs, torsos, and heads. Skillful graphics allowed us to see these often clumsy creatures engage in competitions for the possession of a small cube (see fig. 13; Sims 1994). As they went about their Darwinian wrestling matches, competing for the right to have their constitutive programs reproduced, these boxy critters elicited laughter from the audience. I joined the scientists in their pleasure at these images and experienced the activity of the simulated creatures as cute, especially when they could be interpreted as valiantly failing at their tasks. What made the images funny was a sense that Sims was not fully in control; he had programmed a three-dimensional artificial world (and a visual representation of it) that simulated Newtonian physics, gravity, fluid dynamics, and surface friction, and he had introduced into this world a set of creatures made of

three-dimensional rigid parts that could interact with it. Different creatures had different characteristics and could do more or less well at the
task of capturing the box from competitors. Because the simulated
physics and creatures were programmed together, most behaviors looked
realistic and purposeful. But because Sims occasionally made errors in
modeling physics, sometimes behaviors came off completely wrong, as
when some creatures bounced out of the world because of his mistakes
in modeling gravity. In a brilliant dash of showmanship, Sims showed
videotapes of malfunctioning creatures, explaining that creatures were
"exploiting" bugs in the program and were "making fun of [his] physics."
Sims's ventriloquism delighted the audience and added a sense that his creatures were not only mimicking familiar behaviors but were also mimicking behaviors associated with the playfulness of some life-forms, a playfulness perhaps most readily compared with that of mammalian babies.

Taussig (1993) has argued that mimesis, the ability to copy behaviors, is a faculty often seen as a hallmark of the primitive—as words like
aping and *parroting* attest. And when things considered primitive copy-
cat more advanced behaviors—when dogs dance, birds sing, or apes
sign—we think of them as cute. But things are only cute when they have
relatively little power. When robots mimic behaviors that threaten humans, they are not cute. The cuteness of Artificial Life creatures is produced by *and* produces a sense that they are primitive entities, a sense
that they are capable of miming—perhaps even of parodying or bur-
lesquing—advanced behavior, a sign taken to demonstrate not that they
are not alive but only that they are simpler forms of life.[25] The laughter
at Artificial Life is the spark of life for these simulated creatures. Is it
live, or is it mimesis?

I pause to consider a remarkable image of primitivity in a popular presentation of Artificial Life. The image appeared on the cover of the
Whole Earth Review, a New Age magazine aimed primarily at white yup-
pies. The magazine pictures a naked young black person of indetermi-
nate gender—who nonetheless appears quite feminine (no facial hair,
arms crossed as if to hide her breasts)—who changes from gold to flesh
or from flesh to gold (fig. 14). S/he stands against an emerald stone wall
that is engraved, from left to right, with ancient Egyptian hieroglyphs,
diagrams of molecules, and a schematic rendering of the double helix
appointed with globular gemstones that look as though they were borrowed from an ancient Middle Eastern board game. Tracing the path
from the left of the picture to the right, from the hieroglyphic past to the

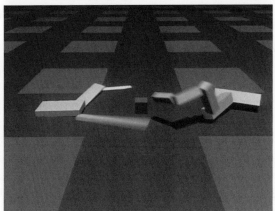

Figure 13. Competing creatures from Sims 1994.
Reproduced with permission.

hydrogen-bonded present, is a gleaming golden circuit diagram of the kind pressed into computer chips. The human figure, positioned against this history of life/science as the history of writing and information, can be seen either as a golden statue coming to life or as a person metamorphosing into a fluid gold organic robot, realizing the Artificial Life philosopher's dream of turning base metals into the gold of life. Given the highly racialized codes through which images are read and produced in American culture, how might we understand this picture of a black person in a popular depiction of Artificial Life, a discipline that popularly pictures its protagonists as white men creating artificial worlds? We could assume that the person pictured is meant to be seen as an Egyptian, arms held in a well-known mummy position. S/he could be coming back to life after prolonged dormancy. Here Egypt is read as an origin point for the social history that has produced Artificial Life.

But the whole affair is more simple and more complicated than this. I read this image against figure 12 (p. 119), the poster for the second conference on Artificial Life. Unlike the man pictured in the poster, the person on the *Whole Earth Review* cover is not rendered as an active subject in the production of artificial life; s/he is an object, marked as primitive by her/his nakedness, youth, femininity, and blackness. And unlike the man in the Artificial Life poster, s/he looks directly at the viewer, and then not defiantly but with a mixture of modesty and allure. Though s/he is probably prepubescent, s/he is quite sexualized, and quite sexualized as feminine and black, a potent combination in a culture in which black women are often represented as hypersexual (see hooks 1992b). Against the straight male lines of the circuit diagram, her curvaceous body is made to give form to the "sexiness" of Artificial Life science. And it does so in a way a white woman's body could not. A white woman's body would read as providing pleasure to a straight white male eye; a black woman's body is readable as "primitive," an object for scientific contemplation, a symbol of organic fertility. If we see the black and golden girl in this symbolic context, she might be seen as a nascent form of artificial life or as a creature ready to birth or care for the property of the godlike and normatively white male Artificial Life scientist.[26]

From these figurings of creation, essence, and primitivity, I now move to the "lives" that Artificial Life organisms lead in their virtual realities. I will escort you through several artificial worlds, weaving this travel around a discussion of a formalism known as the "genetic algorithm," a mechanism that can be used to "evolve" computational processes. The

Figure 14.   Cover of the fall 1992 *Whole Earth Review* (no. 76). Illustration Copyright 1992 by Allison Hershey. Reproduced with permission of artist.

genetic algorithm is a frame on which many Artificial Life systems are built. An examination of its technical and cultural articulations and its use in artificial worlds will enable us to interpret the narrative armatures that allow people to speak of the evolution of computational organisms, families, and populations. It will also give us building blocks for understanding John Holland's Echo system.

THE GENETIC ALGORITHM

The genetic algorithm is a machine learning process invented in the 1970s by the computer scientist John Holland at the University of Michigan (see Holland 1975).[27] Simply stated, the genetic algorithm is a computational procedure that can "evolve" solutions to complex problems by generating "populations" of possible solutions and by treating these solutions metaphorically as individuals that can "mate," "mutate," and "compete" to "survive" and "reproduce." In the mid- and late 1980s, genetic algorithms became popular among computer scientists impatient with the traditional tools of Artificial Intelligence, which were proving inefficient at prosecuting difficult search problems and at solving engineering puzzles requiring multivariate optimization. Many people maintained that if computer scientists wanted to manufacture intelligent behavior, they would do well to look to the organic processes that produced intelligence in the first place. In a move that recalled practices of interdisciplinary borrowing in early cybernetics, computer scientists plundered biology for new ideas and analogies. Researchers building neural nets were inspired by the architecture of the brain. And genetic algorithms were elaborated after the grandest natural process of all: evolution. As David Goldberg, a student of Holland's, phrased it, genetic algorithms were to be "search algorithms based on the mechanics of natural selection and natural genetics" (1989a:1). People working with the algorithm did not entertain the idea that their digital individuals might be alive. But as Artificial Life worlds came to incorporate genetic algorithms as tools for enlivening lineages of virtual organisms, even cautious researchers speculated that computer or robot-based life-forms of the future would probably use genetic algorithms in their reproduction.

The operation of the genetic algorithm is not difficult to understand. A genetic algorithm begins with a randomly generated population of possible solutions to a given problem.[28] Solutions may be predictive procedures, strategies for playing games, or values to be plugged into a complex equation. In Artificial Life systems, solutions often represent virtual creatures and their modes of interacting with the simulated environment and with each other. Solutions are usually encoded as fixed-length bit strings composed of os and 1s, with each string interpreted metaphorically as a "chromosome" that codes for one "individual" in the population. In this formulation, individuals are equal to their "genome" of one unpaired chromosome. The os and 1s making up bit strings are thought of as analogous to the alphabet of four nucleotide bases that form DNA.

Populations of bit strings contain a variety of individuals, some of which are understood to be more "fit" than others. "Fitness" is judged by an individual's ability to perform a given task (in an Artificial Life system this might be to reproduce efficiently with scarce computational resources) or to optimize a preprogrammed "objective function," where "objective" means goal. As Stephanie Forrest puts it, "Each individual is tested empirically in an 'environment' and is assigned a numerical evaluation of its merit by a fitness function F. The environment can be almost anything—another computer simulation, interactions with other individuals in the population, actions in the physical world (by a robot, for example), or a human's subjective judgment" (1993:873). Once the genetic algorithm ascertains initial fitnesses of a bit string population, "parent" strings are chosen to "reproduce," usually in proportion to their observed fitness. Some parents may simply be copied into the next generation, but most are chosen to be in a "mating pool" (Koza (1992:22) and are "mated at random" and in pairs (Goldberg 1989a:12).[29] The algorithm thus features "recombination" where systems like Tierra do not. During "mating," each pair produces two "child" strings by swapping fixed lengths of their chromosomes, as in figure 15, in which bit strings A(1) and A(2) have been chosen as partners. A′(1) and A′(2) are offspring generated from this coupling. (The | indicates the parental site at which code is exchanged.) This process of swapping code, like the biological one that inspires it, is called "recombination," though more often, "crossover."[30] After all crossovers have taken place in a population, "mutations" are occasionally introduced into the "gene pool" by randomly switching some bits from os to 1s and vice versa. Over time, through trial and error and the selection and survival of the fittest strings,[31] the genetic algorithm produces solutions that are ever more effective.[32] Genetic algorithms have found application in

```
parental pair    A (1) = 10101100 | 0101
                 A (2) = 00110110 | 1111

offspring        A' (1) = 101011001111
                 A' (2) = 001101100101
```

Figure 15.   An example of "mating" in the genetic
algorithm.

optimization design for networked manufacturing technologies, pattern recognition protocols, military aircraft, telecommunication networks, and database query tools. Some have even been made commercially available to the general user (see fig. 16). For Artificial Life scientists, the algorithm can be a tool for proliferating pedigrees of program progeny.

THE THEOLOGY OF EVOLUTION

Sherry Turkle (1991) has termed the turn to natural models in computer science a "romantic reaction" to formal, mechanistic, and rationalistic approaches to crafting intelligence. Many people I interviewed did indeed extol what they called the "holistic," "decentralized," and "biological" aspects of genetic algorithms, and one said that the "aesthetics" of the algorithm were more in tune than Artificial Intelligence with "the way nature really built intelligent machines like us." But the desire to transplant the logic of evolution into computers for the purposes of problem solving also derives from a sense that nature is a rationally ordered system, a view not inconsistent with romantic images of nature as possessed of a certain wisdom and tendency toward perfection. For many researchers, nature is an agent that finds solutions to biological design problems. In his book on genetic programming, John Koza writes, "Nature creates highly complex problem-solving entities via evolution" (1992:6). One person I interviewed offered this comparison: "Natural selection is very good at designing complex organisms. Genetic algorithms are modeled on natural selection so they have at least a chance of doing so too." Tom Ray, in a contribution to an early genetic algorithm workshop, asserts that "genetic algorithms exploit the power of evolution to find optimal solutions to problems" (1991:527). David Goldberg, following a tradition of personifying nature as female, offers that nature uses evolution "in her expedient pursuit of betterment" (1989a:309), and he informs us that nature has become quite pragmatic

Artificial Life. . .

Real Solutions.

Welcome To The Future

Something incredible is happening in the computer industry that is commanding the attention of professionals in virtually every field. A new form of software is being born...literally. In an effort to overcome the limitations of traditional programming, developers are exploring technologies which mimic the tried and true methods of Mother Nature.

Digital Darwinism

The most promising of these new technologies is the genetic algorithm (GA). GAs repeatedly generate multiple solutions which adapt, mutate, have offspring, and compete with one another. In this software jungle, only the 'fittest' solutions survive. GAs have been proven effective at tackling resource allocation, distribution, scheduling, budgeting, project management, engineering, and optimization problems of almost every kind.

Genetic Algorithms Grow Up

To address the growing need for customized genetic algorithm applications, Axcélis, Inc. has developed Evolver. *Evolver is the first and most popular commercially available genetic algorithm.* Evolver provides an extendible toolkit of several powerful genetic algorithms for solving numerical, combinatorial, or mixed problems of any complexity. Written as Dynamic Link Libraries under the Windows operating system, users can access and modify the algorithms through their Excel spreadsheet, or their own custom Windows applications.

The successful applications of Evolver have led to coverage and praise in a variety of publications, from Popular Science, to The New York Times. Several universities have implemented Evolver in their computer science curriculums.

If you want to join the hundreds of scientists, engineers, researchers and businesspeople who are using this cutting-edge technology to get more from their computers, call, write, or fax Axcélis for a detailed information packet today.

Figure 16. Ad for Evolver, a genetic algorithm tool kit. Trademark and Copyright Palisade Corporation. Reproduced with permission.

in her computational incarnation: "nature is no spendthrift, nor is she given to whimsy or caprice" (1989a:149). At one Artificial Life conference, Chris Langton emphasized the importance of seeing nature as a kind of experimental scientist: "Nature is our best teacher for engineering; she's had lots of time to work on problems." Here, nature is an optimizing engineer that busies itself (or "herself") with sorting biological blueprints. This view is heir to conceptions of evolution promoted in the genetic reductionisms of sociobiology, in which natural selection is treated as "the agent that molds virtually all of the characteristics of species" (Wilson 1975:67).[33] Through using tools like the genetic algorithm, Artificial Life researchers seek to reproduce exactly this kind of natural selection in computers. As Langton put it at one gathering, "We have to bring about nature in a computer, so that it can be the agent of natural selection."

In these constructions, nature is not just a prudent scientist; it is also a force of progress and improvement, the product of a well-intentioned God. The philosopher James G. Lennox summarizes the position of nineteenth-century natural theology, the outlook from which this view descends: "Animals are structured as they are and behave as they do as a result of being designed for a purpose by a benevolent Creator" (1992:328). Seeing evolution as a force of design and progress directs the way that "fitness" is installed and understood in the genetic algorithm. As the Artificial Life researcher Norman Packard summarizes, "Organisms are replaced by specifications of a device or dynamical rule" and "the fitness function is typically given by an engineering goal" (1989:142). In most genetic algorithms, bit string "fitness" is optimized in prestructured and preexisting environments, and researchers hold that "evaluation functions play the same role in genetic algorithms that the environment plays in natural evolution" (Davis 1996:4). Populations of individuals improve as they solve environmentally given problems.[34]

The biologist Richard Lewontin has argued that the radical separation between organism and environment in much evolutionary biology ignores how the two coexist and co-construct one another, and in so doing reworks a "theological view of a preformed physical world to which organisms [are] fitted" (1984:237). One organismic biologist I interviewed commented that he detected such a theological view in stories of genetic algorithm strings climbing toward higher fitnesses, stories he saw as iconic of the ways neo-Darwinian biology retells salvation tales. Seeing evolution as a process of fitness maximization requires the view that most physical and behavioral traits exist because they have been adaptive, because they have been helpful in allowing organisms to survive and reproduce. It also requires the assumption that such characters have a genetic basis and can therefore be inherited—an assumption easily made in genetic algorithms and Artificial Life, since traits are in fact programmed to be genetic. As the evolutionary biologist Stephen Jay Gould (1979) notes, however, in the actually existing biotic world, biological traits are so interdependent that it makes little sense to talk of particular traits being optimized (see also Dupré 1990). Many people using genetic algorithms would of course concede that genes can have multiple effects and that selection on one trait can be linked to the emergence of other, often deleterious traits. At one genetic algorithm workshop I attended, discussion revolved around lessons researchers might learn from animal husbandry. Breeding, it was argued, should not be a question of maximization or really of optimization, but of compromise to produce improved specimens.

Haraway has noted that the recent focus of evolutionary biology on optimization rather than maximization distances it from natural theological views of nature as perfect. These days, nature takes a long view and finds optimal balances among many variables: "The point of systems design is optimization. Optimization does not mean perfection. A system has to be good enough to survive under given conditions" (Haraway 1991b:64). Koza puts it this way,

> Nature creates structure over time by applying natural selection driven by the fitness of the structure in its environment. Some structures are better than others; however, there is not necessarily any single correct answer. Even if there is, it is rare that the mathematically optimal solution to a problem evolves in nature (although near-optimal solutions that balance several competing considerations are common). Nature maintains and nurtures many inconsistent and contradictory approaches to a given problem. (1992:6–7)

As participants at the genetic algorithm workshop were on the verge of recognizing during their discussion of breeding, the nurturing artificial Nature written into their programs most closely resembles a person engaged in artificial selection. The abstract agency of computational "natural selection" is revealed as the agency of programmers who create and interpret their systems.

Still, many researchers insist that while genetic algorithms explicitly encode a godlike agent of artificial selection, Artificial Life worlds can evolve organisms via fitness specifications that are not preordained.

Fitness is said to be "endogenous," emergent from the logic of the program, from simulated creatures' interactions with one another, not from
the stipulations of the programmer. Tierra is often cited as an example
here, the system that, as Langton once put it, "went all the way," the system in which Ray "removed the hand of God." It is true that the specific strategies that emerge in Tierra are not preprogrammed and that
there is no explicit fitness function against which programs are compared
before they are allowed to reproduce. But there is an implicit fitness function, instantiated in the procedures that allow programs to self-replicate
and that force them to do so in a setting in which there is competition
for resources. It is invested in a programming structure that forces artificial organisms to die: the "reaper." The importance of human agency
in measuring and producing "fitness" is further revealed in a situation
reported in "Artificial Death," in which the evolutionary psychologist
Peter Todd (1993) describes how "immortal" organisms emerged in his
artificial world. Todd reads this as an indication that something is wrong
with his world, instead of taking it as an interesting feature of an alternative reality. Immortals simply do not fit into the evolutionary narrative of progress through (re)productive labor. Artificial Life simulations,
Todd suggests, need population control, some kind of regular death:
"This allows new creatures, with new behaviors, to have access to the
environmental resources they will need to survive, so that constant
turnover of individuals and consequent evolution can take place in the
population" (1993:1048). The worlds in which artificial organisms
evolve, while shaped by narratives of salvation, are well within the
bounds of atheist, antisupernaturalist thought. There is no heaven, no
hell, no limbo. Artificial organisms have no afterlife, although this would
be easy enough to implement. One might even code up a world in which
reincarnation is possible. Instead of reproducing based on their fitnesses,
digital organisms could traverse transmigrational trajectories according
to their computational karma.

   This interest in evolution over individuals—an interest that allows experimenters to terminate creatures in the name of science and to forget
about moments during which they think of them as alive—is made possible by a view of individuals as mere husks for collections of genes.
When genes are the focus of analysis, individual organisms disappear as
interesting units of selection, and often slide away from full identification with their genes. "Fit" genes are distributed throughout a population and may not always aggregate into discrete super-fit creatures.

Sociobiologists first formalized this vision of the world. Committed to a hard Darwinian line that creatures exist because they are directly descended from fit individuals, sociobiologists took as one of their early puzzles the existence of altruistic behavior, helping behavior that potentially lowers an individual's reproductive success. They wondered how such self-compromising behavior could survive the scrutiny of natural selection. In an effort to shore up a competitive image of evolution, sociobiologists argued that genes are the real entities that increase their representation in successive generations (see Hamilton 1964). Genes look after their "extended phenotype" as it is spread across related individuals. Organisms maximize not their individual fitness but their "inclusive fitness"; they act to propagate their genes, whether these are present in themselves or in relatives whose survival and reproduction they can aid. Genes guide the behaviors of individuals and of groups (hence *sociobiology* as the proper name for the study of genes' effects) and are ultimately of greater importance than either. In genetic algorithms, a similar logic obtains. Patterns of os and 1s occurring in individual strings are often of more interest than individual solutions. Koza writes,

> In the genetic algorithm, as in nature, the individuals actually present in the population are of secondary importance to the evolutionary process. In nature, if a particular individual survives to the age of reproduction and actually reproduces sexually, at least some of the chromosomes of that individual are preserved in the chromosomes of its offspring in the next generation of the population. . . . It is the genetic profile of the population as a whole . . . , as contained in the chromosomes of the individuals of the population, that is of primary importance. The individuals in the population are merely the vehicles for collectively transmitting a genetic profile and the guinea pigs for testing fitness in the environment. (1992:37)

If evolution is an abstract process of information replication, then the vehicles through which it occurs are just so much fluff. On this view, genes provide a kind of immortality—as do other self-replicating information structures that steer the behavior of organisms, like "memes," Dawkins's term for ideas that travel from one mind to another. The focus on genes and memes acquires an odd spiritual torque in Tom Ray's argument that Artificial Life is not a sublimation of a "religious" desire to achieve immortality:

> I prefer to achieve immortality in the old-fashioned organic evolutionary way, through my children. I hope to die in my patch of Costa Rican rain forest, surrounded by many thousands of wet and squishy species, and leave

it all to my daughter. Let them set my body out in the jungle to be recycled into the ecosystem by the scavengers and decomposers. I will live on through the rain forest I preserved, the ongoing life in the ecosystem into which my material self is recycled, the memes spawned by my scientific works, and the genes in the daughter that my wife and I created. (1994a:204)

In contrast to the monogenetic God Ray played in Tierra, here he is co-shareholder in the genetic endowment of his child. He becomes part of a terrestrial genetic algorithm that offers beatific vision.

REPRODUCTION, SEX/GENDER, AND THE HETEROSEXUAL FAMILY

Reproductive dynamics in genetic algorithms are inspired by processes of reproductive coupling in populations of sexually reproducing animals with sequestered germ cell lineages (even as bit strings equal single un-paired "chromosomes" [are haploid] where many sexual animals carry paired chromosomes [are diploid]).[35] Humans are a familiar example of such animals, and for this reason, culturally built notions of gender, sexuality, and family frequently wind their way into artificial worlds. Affairs might be different if asexual plants were the model of choice or if polyploid sexual organisms like maize were the preferred casts for *life-as-it-could-be.*

In the standard genetic algorithm, strings pair up to produce offspring using the procedure called crossover, and the terms "parents" and "children" are routinely used to refer to such strings. Lawrence Davis writes, "In nature, crossover occurs when two parents exchange parts of their corresponding chromosomes. In a genetic algorithm, crossover recombines the genetic material in two parent chromosomes to make two children" (1996:16). In a permutation of the genetic algorithm called genetic programming, parents consist of programs that trade subroutines (see fig. 17). Artificial Life worlds like Yaeger's PolyWorld offer us bit string "genes" that code for virtual creatures' "physiology" (their size, speed in their simulated world, etc.) and for mini-programs that endow creatures with a "neural architecture" (allowing them to navigate, find "food," etc.).[36] Genes standing for these items recombine as they are transmitted from parental dyads to offspring (Yaeger 1994).

There are a number of ways we might understand the exchange of bits between strings, but the metaphor of productive heterosex is glee-fully emphasized by most authors. Goldberg writes, "With an active pool

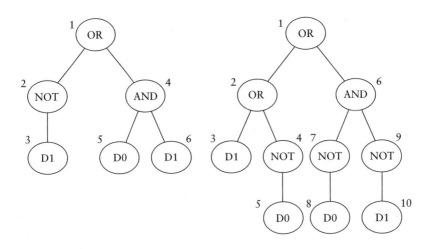

Figure 17.   Two parental programs from John Koza, *Genetic Programming: On the Programming of Computers by Means of Natural Selection*. Copyright 1992 MIT Press. Reproduced with permission.

of strings looking for mates, simple crossover happens in two steps: (1) strings are mated randomly, using coin tosses to pair off the happy couples, and (2) mated string couples cross over, using coin tosses to select the crossing sites" (1989a:16). Levy's *Artificial Life* uses more fanciful language: "Next, the strings mated. In a mass marriage ceremony worthy of Rev. Moon, each string was randomly paired with another" (1992b:163). At a talk at SFI one day, I heard a notable genetic algorithmist say that he thought intuitively about crossover by "thinking about what it means to recombine my genes and my wife's genes." In these descriptions, monogamous heterosexual marriage (even if random) is considered a realistic template for natural processes of sexual coupling for reproduction.

This heterosexual grid cannot contain the fact that there is something odd about the way bit strings "reproduce," and this is that they enact sexual reproduction with absolutely no sexual difference (reflecting on this, one person told me that genetic algorithms are actually "more like bacteria sex than human sex," an interpretation that does not keep researchers from speaking of families, "incest taboos," and "marriage restrictions," as I discuss below). Like the artificial organisms for which they often code or with which they are often conflated, bit strings never menstruate, ovulate, get pregnant, ejaculate, or have problems maintaining erections. The idea that mating can happen between structurally identical entities recalls what Keller has called the masculine bias of mathematical population genetics. In this discourse, all individuals are structurally equal, all just bags of genes. Keller writes, "Effectively bypassed with this representation were all the problems entailed by sexual difference, by the contingencies of mating and fertilization that re-sult from the finitude of actual populations and also, simultaneously, all the ambiguities of the term reproduction as applied to organisms that neither make copies of themselves nor reproduce by themselves" (1992d:132).

"Sex" becomes an abstract affair. Likewise in the genetic algorithm: in the "sexual operation" of crossover (Koza 1992:101), no disorderly bodies intervene, and artificial organisms become masculinized seed-strings. The "essence" of sexual reproduction—indeed of "sex"—is distilled into an out-of-body experience. As one male Artificial Life scientist summarized this notion for me in another context, "It doesn't make any difference whether people make babies in a test tube or by fucking."[37] And as another researcher remarked when I asked him about the absence of development, embryogenesis, and gestation in most Artificial Life models, "Pregnancy is merely an implementation problem." As in many other cultural sites, idealized male physiology is taken as normative, and female physiology is seen as deficient or as laden with supplementary features. Artificial worlds employing the basic genetic algorithm encode these prejudices, as Helena reminded me in one interview:

> There's one side of me that says it's really strange how they leave out sexes when they have reproducing things in these models. There's another side of me that says, well, that's not all that biologically unrealistic; they're just talking about very simple organisms. But, on the other hand, it does seem like there are a lot of dynamics that go along with that. When they talk about the organisms in their simulations as "these guys," there's a little part of me that says, well, they're basically creating these all-male worlds.

Against the grain of this generalization, there exist several systems that attempt to take sexual difference into account. Some are devoted to modeling populations of real-world organisms (e.g., Taylor et al. 1989; Terzopoulos, Tu, and Grzeszczuk 1994); others include difference in the service of exploring theoretical issues like sexual selection (e.g., Todd and Miller 1991, 1997; Werner and Dyer 1992; Miller and Todd 1993; Werner 1997). In systems like these, sexual difference is often simply coded into genes for virtual organisms, along with gendered stereotypes of males as active and females as passive.[38] In Gregory Werner and Michael Dyer's system, for example, which explores the evolution of signals in mate choice, simulated "animals" existing in a grid-space come in two flavors: females, who sit still, and males, who search for them: "When a male finds a female (moves onto the same grid location that she is on), the animals mate and produce two off-spring, a male and a female. The parents' genome, which encode their neural network brains, are combined using the standard genetic operations of crossover and mutation to produce the genome of the off-spring" (1992:663). SimLife also offers genetically determined sexes in its tool kit for modeling real and possible life-forms. Here, "female" genomes contain specifications for "gestation," which can affect the amount of "food" a female needs to reproduce, the number of creatures she can produce, and the number of time steps reproduction takes. Here, a data structure representing a female can contain another data structure that represents the result of recombining the female's "genome" with the "genome" of a male. The "organisms" coded by these nested data structures are released into the world when enough time has passed and when the "female" "carrying" them has gathered enough "energy." When offspring are born, they take with them a percentage of the "mother's" energy and inherit her "health" level. "Mutagens" in the outside environment can also cause the gestating organism's genome to change during gestation (note that this is not an effect of gestation as such; just as in "life begins at conception" models, the gestating creature is considered an entity separate from its mother). For all this complexity, "pregnancy" does not change the fact that virtual creatures pop into the world completely determined by their genes. Another system called the Sugarscape runs with genetically coded sexes, adding the requirement that, to mate, "agents" must be "fertile"—though this specification refers to both sexes and does not lead to a SimLife-like caricature of pregnancy.

In all these models, "sex difference" is genetic (and creatures are haploid, even as they are fashioned to remind us of diploid creatures).[39] In simulations purporting to stencil real and possible complex sexed behaviors, such genetic determinism needs to be questioned. Even in the existing world, genes do a poor job of pinning animals to sexes, which materialize as never quite clean alignments of genitals, chromosomes, gonads, hormones, and behavior (see Fausto-Sterling 1992). And this only describes some animals; definitions need to be modified to explain sequentially hermaphroditic fish and parthenogenetic lizards, for example, to say nothing of polymorphously sexual plants.

What supports the fixation on binary sex difference as genetic? In part, it is a heterosexual faith that sexual difference must have evolved to facilitate recombinative reproduction, and as evolved must be under genetic control. But while there are connections between sexual difference and procreation, sexual activity comes in many varieties, not all of which are about reproduction. Heterosexuality as an identity or normative practice need not logically exist for evolution in sexual organisms to occur (and it has not, historically). Perhaps one of the roots of confusion is that many of us conflate sex as an activity with sex as a biological identity. The idea that one can "have sex" as well as "have a sex" is a trick that fools us into believing that having a sex in contrast to another's sex is a necessary prerequisite for sexual activity. As the feminist philosopher Monique Wittig puts it, "The category of sex is the political category that founds society as heterosexual. As such it does not concern being but relationships (for women and men are the result of relationships), although the two aspects are always confused when they are discussed" (1992:5).[40]

Regardless of whether sexual difference is installed in genetic algorithm or Artificial Life systems, why is the mechanism of crossover so important? Why isn't replication with mutation, as in Tierra, sufficient to breed a variety of individuals? Doesn't the crossover operation scramble potentially useful genetic combinations? On the contrary, researchers argue, it generates them; the recombination of substrings allows a population to produce many different kinds of individuals quickly: "This mixing allows creatures to evolve much more rapidly than they would if each offspring simply contained a copy of the genes of a single parent, modified occasionally by mutation" (Holland 1992b:66). Sex as recombination is seen as a source of innovation that allows species to be adaptively flexible; far from fracturing the integrity of the "seed," it enriches it.

This view is brought to us by a conception of sex as a system for producing unique and ever-improving individuals. Koza writes, "The crossover operation produces two offspring. The two offspring are usually different from their two parents and different from each other. Each offspring contains some genetic material from each of its parents" (1992:23). Holland adds, "The algorithm favors the fittest strings as parents, and so above average strings (which fall in target regions) will have more offspring in the next generation" (1992b:68). Davis puts it all together: "If all goes well throughout [the] process of simulated evolution, an initial population of unexceptional chromosomes will improve as parents are replaced by better and better children. The best individual in the final population produced can be a highly evolved solution to the problem" (1996:5).

The idea that children will be different from and better off than their parents is premised on an understanding of kinship as a system that continually generates future possibilities. This is a particularly middle-class (and class-conscious and -climbing) view primarily associated with Europeans and Americans. As Strathern writes for the English context, in terms that can extend to describe the common sense of many of the people I interviewed, "Kinship delineate[s] a developmental process that guarantee[s] diversity, the individuality of persons and the generation of future possibilities" (1992a:39). "Increased variation and differentiation invariably lie ahead, a fragmented future as compared with the communal past. To be new is to be different" (Strathern 1992a:21). In this kin system, children are "new" "individuals" that emerge from parental relations. They are not reincarnations of ancestors, spitting images of one parent, or members of a class of young people that resemble each other more than any resemble their parents—all alternative ways that children have been construed in world cultures. Strathern writes of the notion of unique children what one might write of the brave new organisms of Artificial Life and genetic algorithms:

> The child's guarantee of individuality lies in genetic origin: its characteristics are the outcome of a chance combination from a range of possibilities. . . . Genetic potential . . . maintains an array of possible characteristics from which an entity might emerge; the future is known . . . by its unpredictability, and one would not necessarily wish to anticipate it. (1992b:172)

The figuring of kinship through the language of genetics is of recent vintage—though, in many ways, culturally particular notions of kinship have been wired into evolutionary formulas for a long time. According to Strathern, when Darwin wrote *On the Origin of Species,* he sketched evolutionary relatedness after analogies to human kinship:

> Darwin drew on the prevailing ideas of his time concerning genealogy and relatedness between human beings in order to depict degrees of affinity between other species. In the twentieth century Euro-Americans have turned this back on itself, and conceive biological relatedness as primordial and prior to the constructs human beings build upon it. (1992b:16)

This conception has shaped how Euro-Americans think of kinship as concerned with the social organization of "the facts of life,"[41] as a social arrangement modeled after and attentive to biogenetic connection— rather than as a concept that refers to how people make sense of social connection in ways that may make reference to biogenetics but that may also implicate political, class, caste, racialized, sexualized, and religious affiliations (see Geertz and Geertz 1975; Collier and Yanagisako 1987; Weston 1991). "Relative" has become, for Europeans, Americans, and many others in the sweep of science, a biological term. For this reason, kinship terms from the Euro-American lexicon have been read onto bio-genetic connections and then used to structure knowledge about bio-genetic categories themselves. One genetic algorithmist I spoke with did not stop at "parents" and "children" in describing relationships between bit strings but added terms like "grandparent," "aunt," "cousin," and "husband," folding these kin terms into the very way his algorithm, which he understood to mirror "nature," operated.

Some genetic algorithm researchers have suggested that strings might be prevented from crossing with strings too similar to themselves—an operation that could stall the generation of new solutions—and have introduced into their programs an "incest taboo" (Todd and Miller 1991; see also Eshelman and Schaffer 1991, 1993). This restriction seeks to imitate in an artificial medium a presumed natural mechanism for pre-venting mixing between genomes similar by relatedness ("inbreeding"), but in using the term "incest taboo" it invites the conflation of this mech-anism with what are in fact culturally constituted rules governing the separation of cognatic and conjugal roles in Euro-American families— roles determined by age, gender, and generation. Most Artificial Life "in-cest taboos" only gesture toward such elaborations, however, simply pre-venting similar strings from mating and not necessarily blocking the recombination of strings "related" by descent.

In Western kinship constructs, the act of heterosexual intercourse that "produces" children is thought to be the generative knot that produces "families" and makes people "related" (Schneider 1968). In genetic al-gorithms, the relatedness of digital organisms is originated through cou-plings fashioned exactly after this model. The anthropologist David Schneider's reflections on the illustrative middle-class American kinship case are directly relevant here: "In American cultural conception, kinship is defined as biogenetic. This definition says that kinship is whatever the biogenetic relationship is. If science discovers new facts about biogenetic relationship, then that is what kinship is and was all along" (1968:23; see Yanagisako 1978 for an argument about the Euro-American specificity of Schneider's "America"). Schneider underplays the extent to which biogenetic categories are already inflected by the kinship categories that call on them, but he is right that biogenetics has become the preferred idiom for thinking about kinship. And in an age in which genetics has become an information science, kinship is becoming informatic. It should be no surprise that information structures that spawn other information structures can be described as "parents."

What is preserved in biogenetic/informatic logics like these is the sense that family ties are about descent lines. One genetic algorithm researcher told me that he would "define a family as parents and children, without going laterally." Another said that being related to someone meant that "you have common descent." What these formulations share is a focus on the nuclear family and on the parent-child link, a link that privileges the constant reproduction of future relations and individuals and is of a piece with the way genetic algorithms generate futures full of new solution strings. Koza hooks this in with a discussion of population tabulations in genetic algorithms, highlighting a theme of progress through rational procreation, outfitting artificial natural selection as a sort of bureaucrat: "A genealogical audit trail can provide . . . insight into why the genetic algorithm works. . . . [T]wo parents were selected to be in the mating pool in a probabilistic manner on the basis of their fitness. . . . [T]hey then came together to participate in crossover. Each of the offspring produced contained chromosomal material from both parents. In this instance, one of the offspring was fitter than either of its two parents" (1992:25). The oddly Euro-American cast of this kinship imagining (bolstered in formulas in which bit string couples have the ideal two children of the middle-class household) is illustrated in Goldberg's proposal for running several genetic algorithms in parallel. He suggests that the algorithm be mapped onto a community structured according to the schematic diagram pictured in figure 18, explained as follows:

> Here the genetic algorithm is mapped to a set of interconnected communities. The communities consist of a set of homes connected to the centralized, interconnected towns. Parents give birth to offspring in their homes and perform function evaluations there. The children are sent on to a centralized singles bar (in town) where they meet up with prospective mates. After mating, the couples go to the town's real estate broker to find a home. Homes are auctioned off to competing couples. If the town is

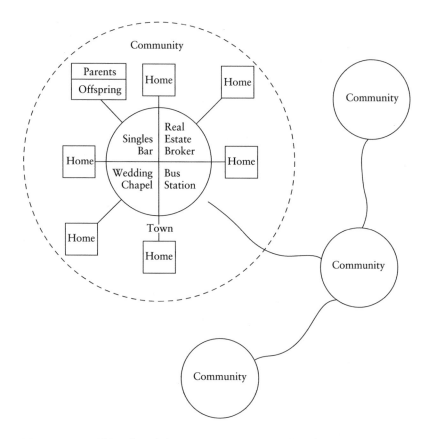

Figure 18.   "Object-based design, a community model of a GA." From
David Goldberg, *Genetic Algorithms in Search, Optimization, and Machine
Learning* (fig. 5.37). Copyright 1989 Addison-Wesley Publishing Company,
Inc. Reprinted by permission of Addison-Wesley Longman, Inc.

> currently crowded, the couples may also consult a broker about homes in
> other communities, and if necessary, they may go to the bus station to
> move to another community. (1989a:210)

Again, in a formulation that combines the parodic and the programmatic,
a culturally particular kinship system is folded into a computational formalism that purports to mime biological reality.

ELEMENTARY STRUCTURES
OF ELECTRONIC KINSHIP AND RACE

If stories of kinship and family in genetic algorithms often encode
images of communities characteristic of Euro-America, they are also in-

habited by the racialized logics that produce these systems. A detour into a technical point illustrates how this can be so. Some researchers have argued that a persistent problem in the genetic algorithm is convergence to suboptimal solutions: "In optimization, when the GA [genetic algorithm] fails to find the global optimum, the problem is often attributed to premature convergence, which means that the sampling process converged on a local rather than the global optimum" (Smith, Forrest, and Perelson 1992:4). To remedy this problem, some have proposed regulating the kinds of strings that can cross, through introducing different sorts of "mating restrictions," like the "incest taboo" above, in which similar strings are prevented from recombining. Strings can also be prevented from mating with strings too different from themselves (see fig. 19), a process one researcher has called "marriage restriction" (Booker 1985; see also Goldberg 1989a). The reasoning is that crosses between strings that are too different might disrupt a population's accumulation of useful and potentially optimal genetic combinations. In discussions of this latter strategy, researchers often employ highly racialized imagery. In Yaeger's PolyWorld, for example, restrictions can be enforced to encourage the divergence of populations (as genetically interbreeding groups) using a tool called the " 'miscegenation function' (so dubbed by Richard Dawkins), that may be used to probabilistically influence the likelihood of genetically dissimilar organisms producing viable offspring; the greater the dissimilarity, the lower the probability of their successfully reproducing" (Yaeger 1994:272).

"Miscegenation" is of course a loaded term, and refers not to mixing between species or incipient species but to mixing between "races." *Webster's Third New International Dictionary* provides the following definition: "a mixture of races; *esp* : marriage or cohabitation between a white person and a member of another race." Miscegenation has been a word with violent connotations; its most frequent use has been in U.S. "antimiscegenation laws," laws written to prevent marriages between "whites" and "nonwhites." The racial and eugenic logics skittering below the surface of genetic algorithms are made explicit here and key us into a notion of "races" as distinct genetic groups, rather than as socially constructed groupings.[42] Genetic difference, coded here as biological race, is to be handled carefully, with populations kept pure of contamination from others. In the universes of Artificial Life, sexual recombination, which produces new combinations of traits, must be kept within boundaries, lest lineages lose their vigor. The shadow of eugenics haunts

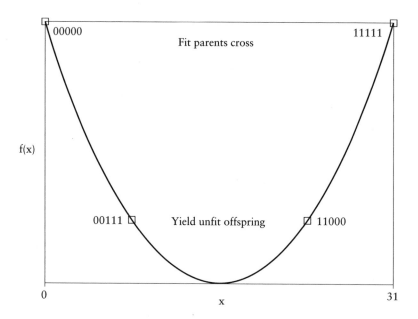

Figure 19. "Simple bimodal function illustrating need for mating restriction. Crosses between dissimilar near-optima almost always cause lethals." From David Goldberg, *Genetic Algorithms in Search, Optimization, and Machine Learning* (fig. 5.23). Copyright 1989 Addison-Wesley Publishing Company, Inc. Reprinted by permission of Addision-Wesley Longman, Inc.

selection stories in genetic algorithms. One writer notes that the algorithm is "more like 'artificial selection,' and is perhaps akin to animal breeding—or a kind of silicon version of eugenics" (Jenks 1994). Another summons up the imagery of eugenics even as he tries to banish it. In a discussion of a system in which genetic algorithms code for stock market trading strategies, Gibbons Burke writes, "If this sounds hauntingly like some Hitlerian eugenics nightmare, don't fret. No blood will be shed because these traders, this entire market, exist only in the memory of a NeXT computer at the Santa Fe Institute in New Mexico" (1993:26). Haraway has suggested that "racial hygiene and its typological syntax are not supported by genome discourse, or by artificial life discourses in general" (1997:248). As these examples show, this is not entirely true; early-twentieth-century notions of races as distinct populations can still shape the biology of bits and bytes.

Racial themes are not only activated in moments of colorful explanation of genetic algorithms; they are also frequently articulated through the language of biogenetic kinship. Not only is biogenetic kinship a peculiarly Euro-American formula, it is one that has supported a rhetoric of purity for the category "white" in U.S. history (and as it has been imposed on and borrowed by other peoples, for other racialized groups as well). The biogenetic model is inhabited by the bilateral and fractional genealogical logics of Western kinship, even as it purports to speak the universal and basic language of biology. A glance at an artificial world called the Sugarscape will illustrate how this legacy works and how it might tint the informatic kinship of Artificial Life systems.

The Sugarscape, developed by the complexity researchers Robert Axtell and Joshua Epstein of both the Brookings Institution and the Santa Fe Institute, is an artificial world in which "agents" interact in an environment filled with a "resource" called sugar (see Epstein and Axtell 1996). Agents act according to rules encoded in a genome, and their local behaviors lead to high-level aggregations, visualized on computer screens as clusters of differently colored dots. The Sugarscape, which was frequently referred to as "Artificial Social Life" when I was at SFI, is a tool for simulating the emergence of elementary social relations from the interactions of individuals. Epstein noted in one interview, "We have . . . grown entire little Proto-histories of society, in which cultural groups— Reds and Blues—emerge from a primordial 'soup,' and migrate to separate sugar peaks, where populations grow, forcing a diffusion back down into the lowland between the sugar mountains, where combat, and cultural assimilation . . . perpetually unfold" (quoted in Little 1993:17). Epstein and Axtell, in their talks and in their writings, routinely refer to the "cultural groups" in Artificial Social Life as "tribes" or "tribal formations," understood as "culturally homogeneous" entities marked by distinct colorings.

Agents produce new agents as they "mate" with others of opposite "sexes," using a genetic algorithm. The Sugarscape is capable of producing "family trees" that record lineages generated by these couplings (Epstein and Axtell 1996:9). When this was explained in talks, people would often wonder aloud about the "marriage arrangements" of agents. One man asked whether, with certain parameters turned on for "sex," "harems" might emerge. Though humorously posed, it was a serious question. It was also a question that painted agents as potentially "primitive," insofar as nonmonogamous marriage patterns are associated in the white imagination with degenerate, often Orientalized others. One person gave all this a biblical twist when he said, "You can run populations in possible worlds—see why the guys that don't have the right command-ments lost out—they were busy chasing each others' wives around." (As it turns out, agents may have multiple couplings, but there is no restriction on which "sex" may accumulate these.)

While this language of "tribes" and "harems" plugs us into a white racial imagination, so also do the kinship logics that fashion such ag-gregations. This is perhaps best demonstrated by setting this simulation of simple social relations alongside a seemingly distant practice: U.S. government designations of what it means to be an "authentic" Native American. The same definitions have been used to classify Native Amer-icans as are used in this program to grant agents their initial "tribal" af-filiation. In discussions of the Sugarscape, there is an assumption that cultural traits as well as biological traits are inherited bilaterally. At the moment that agents reproduce using a genetic algorithm, they also en-dow their offspring with a "cultural" identity (like "Red" or "Blue"), the result of matching parental identities, or if parents do not match, of giving the process over to Mendelian inheritance (often making the off-spring half-and-half, with a kind of dominance determining final iden-tity). These "cultural" identities, like agents' "genes," are made of os and 1s. When SFI audiences of the Sugarscape say things like "let's call this group the Navajo" or suggest that the system can be used to model the ancient Anasazi (Epstein and Axtell 1996:164), this is more than carefree analogy making. It is suggested by the very logics of kin and cul-tural belonging installed in the program—logics that I argue have also been enacted in the way the U.S. government has determined Native American identity in line with Euro-American kinship, thereby forcing Euro-American grids onto diverse systems while at the same time main-taining that these grids map onto primordial biogenetic truths.

When the U.S. government entered into treaties with indigenous peo-ples in North America—treaties drafted to contain Native Americans in reservations—it committed itself to providing various services to in-digenous populations. These services, including health care and educa-tion, were to be given to "legitimate" members of Native American "na-tions." In 1887, a piece of congressional legislation known as the General Allotment Act dissolved most tribal landholdings in favor of deeding land to individual Native Americans, a move that changed the ways le-gitimate identity was figured. From this point forward, the U.S. govern-ment used an individual's "blood quantum," or fraction of Native Amer-ican parentage, as the measure of his or her cultural belonging. Given the fact that laws against white–Native American intermarriage were inconsistently enforced, these fractions could be safely assumed to get smaller and smaller until fewer people fit the rather strict federal definition and the government's obligations were reduced to a vanishing minimum (Jaimes 1992). What these definitions did, aside from set the stage for ethnocide through legal means, was assume that culture equals kinship and that kinship is a self-evident matter of hereditary relationship, or "blood quantum," a formulation that later would be said to refer not to blood, really, but to biogenetics. Thus formulations for designating Native American belonging that were forged in part for an explicitly political project were also thought to be "natural." This definition assumed that kinship as both a biological and cultural category undergirds social relations in the most obvious way. The biogenetic grid was taken as the basis for all families everywhere and for cultural aggregations above the family level as well. Native American identity was constructed by the government by these logics, even though the hundreds of native groups have very different ways of understanding family, kinship, and belonging. The white American definition of kinship was read directly onto the nature of elementary social relations.[43]

"Blood quantum" rubrics of belonging for Native Americans have stayed on the books far longer than those for any other group; indeed if one wants to buy "authentic" Native American art in Santa Fe, as many Institute visitors do, one needs to make sure that the art is tagged with a certificate of authenticity that records the artist's census number, issued by the government in accordance with blood quantum laws.[44] The point I want to make is that definitions of authentic Native American belonging by the U.S. government converge with the kinship systems of the white scientists at the Santa Fe Institute and direct how the Sugarscape is programmed and interpreted. In this simulation model of "tribes," cultural belonging derives from parents. We might speak of agents as claiming membership in tribes by "information quantum" (though, to be accurate, fractional inheritance, while installed in the program, is often covered over by dominancelike effects). Like Albert Einstein wearing a headdress in the Institute's stairwell photograph, Sugarscape agents are dressed up as Indians, but they also stand as translucent representatives of scientific and Euro-American logic.

The Sugarscape is a bit more sophisticated than this analysis would suggest, however: cultural belonging, while directly and bilaterally inherited from parents, can change during agents' lifetimes through interaction with other agents; it is a parameter that, unlike "sex," is not encoded into an unchangeable genome. And parents may pass on an acquired cultural belonging, adding a kind of Larmarckian inheritance to the system. Epstein and Axtell thus afford the possibility for "cultural assimilation," for agents to change "tribes" during their lifetimes. This wrinkle makes the Sugarscape into a kind of computational melting pot—an image Epstein and Axtell actually evoke when they give an example of how agents may belong to the same culture (i.e., be "Blue" or "Red," based on a similar preponderance of os or 1s, respectively), even as they may differ within that "culture" (i.e., express os and 1s at different locations in their "cultural" bit strings): "two individuals might consider themselves 'American' culturally, while differing politically, religiously, or in other respects. An interesting feature of this agent group membership rule is that agents can be very different culturally, measured position-by-position and yet be members of the same group" (Epstein and Axtell 1996:77–78). "Culture," seen this way, is decoupled from biology, but in a manner that then leaves the logics of Euro-American kinship sedimented into elementary kin connection ("culture" also acquires its own kind of white raciality, styled as it is after a melting pot logic that in the United States has been a code for coercing assimilation to a white norm). The Sugarscape thus speaks in a forked tongue; conflating cultural and biogenetic belonging at moments of inheritance but then separating them to ensure that biology is not identified with culture. Even so, the Sugarscape offers many scenarios in which "culture" and "biogenetics" reconverge: "The fundamental drive for sugar produces migration to one or another of the sugar peaks, and thus spatial segregation into two subpopulations in which mating and cultural transmission occur. And each sub-population converges (culturally) to pure Red or pure Blue; the tribes are formed. Sexual reproduction now increases each tribe's population" (Epstein and Axtell 1996:92). Sugarscape tribes come to be "pure," and purely identified with biogenetic kin groups, just like legal constructions of Native American tribes.

This appeal to white kinship categories as natural underwrote events at a field trip I took with SFI researchers and staff to a Tewa archaeological site, where I heard an archaeological presentation on the social organization of ancient Tewa villages. The white American archaeologist, in an effort to give an objective picture of Tewa household organization, translated all Tewa kin terms into Euro-American terms, a move that eased "our" comprehension of a different way of life (all members of the audience were white) but also presumed that our categories were less paved over with cultural presuppositions. Like the Sugarscape program, this approach conflates Euro-American and scientific kin terms, consolidating an image of both science and white ethnicity as unmarked, normative, and "acultural" cultural spaces.

In her study of particle physicists, the anthropologist Sharon Traweek characterized the world of the scientists she worked among as "an extreme culture of objectivity: a culture of no culture" (1988:162). In her ethnography of white women, Ruth Frankenberg (following the work of Trinh Minh-ha [1986–1987] and Chandra Mohanty [1984]) examined how the women she interviewed often understood their whiteness as an "apparently empty cultural space" (1993:192). White people often imagine that their categories are not "cultural"; that "ethnicity" and "race" are markers of other people, not them. Frankenberg holds that a view of white culture as no culture serves to obscure the real power of whiteness as an unspoken metric against which "others" are judged (see also Baldwin 1984; hooks 1992a; Roediger 1994; Haney López 1996). The same might be said of science. The power of science to represent itself as objective relies on the production of a set of bounded nonscientific "others": the social, the irrational, the primitive, the religious, the emotional. I suggest that there are points of contact between the construction of whiteness and science as cultures of no culture. The mechanisms for producing whiteness and science are not only conceptually similar but are also implicated in one another practically and historically.[45] When the authors (and audience) of the Sugarscape and the white archaeologist use Euro-American kinship terms as scientific terms, they take white culture as a lens through which they can transparently see objective reality. The legal scholar Barbara Flagg notes, "The most striking characteristic of whites' consciousness of whiteness is that most of the time we don't have any. I call this the *transparency* phenomenon: the tendency of whites not to think about whiteness or about norms, behaviors, experiences, or perspectives that are white-specific" (quoted in Haney López 1996:22). The "dislocating dazzle of 'whiteness'" (Gilroy 1993:9) blinds many Artificial Life scientists to the ways their images of kinship and primitivity are cultural constructions built on the foundations of a white cultural imagination. In many ways, the genetic algorithm is animated by this imagination, and the worlds it supports resonate with its logic. Which brings us to the final exhibit, John Holland's Echo system.

ECHO

The Echo system is a computational platform for simulating evolutionary processes in a variety of "complex adaptive systems," from immune systems to ecologies to economies (see Holland 1992a, 1993, 1994, 1995; Jones and Forrest 1993). It is an excellent example of a model produced in the service of the sciences of complexity, since it contains groups of "agents" whose "interactions" give rise to nonlinear dynamics. In the sociologist Joan Fujimura's (1996) terms, it is a kind of theory-methods package, an instrument in which scientific assumptions and techniques come bundled together. "Echo system," of course, plays on "ecosystem," suggesting that the platform can "echo" processes observable in natural systems. Stephanie Forrest compares Echo to model organisms in biology, creatures used to emulate aspects of other organisms' behaviors. Holland describes Echo as well as some of its possible uses:

> Echo provides for the study of populations of evolving, reproducing agents distributed over a geography with different inputs of renewable resources at various sites. . . . Collections of agents can exhibit analogues of a diverse range of phenomena, including ecological phenomena (e.g., mimicry and biological arms races), immune system responses (e.g., interactions conditioned on identification), evolution of metazoans (e.g., emergent hierarchical organization), and economic phenomena (e.g., trading complexes and the evolution of "money"). (1992a:186)

Though meant as a general purpose simulator, Echo is a frequent reference in Artificial Life, where its status as an artificial world capable of modeling evolutionary dynamics brings it into dialogue with systems like Tierra, SimLife, and PolyWorld. Unlike the authors of these simulations, Holland does not entertain the notion that agents in Echo are alive, though he describes them in ways that suggest they have will and intention. Holland thinks of Echo not as an alternative universe but as an apparatus for computer-aided thought experiments. Such experiments do not necessarily capture the complexity of real systems, nor are they meant to replicate data sets from real world experiments; rather, they offer partial perspectives on how complex adaptive systems work. Holland compares simulation to political cartooning: "The modeler (cartoonist) must decide which features to make salient (exaggerate), and which features to eliminate (avoid) in order to answer the questions (make the political point)" (Holland 1995:11). Modeling decisions, as Holland's comparison suggests, often contain cultural and political commitments.

Because Holland is rarely at SFI (he spends most of his time at Michigan), much of my practical understanding of Echo comes from conversations with Terry Jones, a computer science graduate student who, during my stay, was charged with implementing and developing the system.[46] Jones was frequently called on to explain Echo in workshops and to visitors. Near the end of my fieldwork, Echo was showcased to Vice President Al Gore, who was interested in the use of simulation in ecological modeling. Echo remains one of the Institute's most publicized projects.

In the Echo universe, strings of alphabetic characters are arrayed in structures manipulated, moved, and transformed by the program making up and running the system. In Holland's favored description, strings are the genomes of "agents" that interact in an environment filled with "renewable resources." Holland picks out the proper interpretation: "A precise description of Echo begins with definition of the individual agents. The capacities of an agent are completely determined by a small set of strings, the 'chromosomes,' defined over a small finite alphabet" (1992a:186). Agents "interact" when the program calls different agent structures into association. Interactions fall into three categories: "combat," "trade," and "mating."

Agents are located at "sites" on a rectangular array, and their most important task is collecting resources. As an agent remains on a site, the site exacts a "tax" the agent must pay with resources stored in its "reservoir." In an article in the SFI *Bulletin*, Terry Jones notes, "You can't stay alive without expending energy at some rate . . . And, if an agent gets charged a tax he can't pay, he is considered bankrupt and essentially is taken out of the world" (quoted in Little 1993:15). Agents acquire new resources from other agents in "combat" or "trade," or by "migrating" to new sites. Echo agents are made of the same letters (a's, b's, c's, and d's) that designate the resources in this computational land, and so, in collecting resources, they maintain themselves and can also be considered to be gathering materials to construct offspring. Substrings of the letters of which agents are constituted are grouped into "tag" and "condition" "chromosomes" that determine how agents will interact, that is, with whom they will fight, trade, or mate. Agents are brought into interaction in pairs, and tags and conditions are compared to see what sort of action will occur. Agents are tested first for combat, next for trade, and finally for mating. Holland has

```
if (AGENT_OFFENSE_TAG_LEN (a2) >= AGENT_COMBAT_COND_LEN (a1) &&
    !memcmp (AGENT_COMBAT_COND (a1), AGENT_OFFENSE_TAG (a2),
    AGENT_COMBAT_COND_LEN (a1)))){

    /* Combat will happen */
}
else {
    /* Combat will not happen */

}
```

Figure 20.   Echo code that tests for combat.

written of tags, "It is convenient to think of the tags as displayed on the exterior of the agent, counterparts of the signature groups of an antigen or the trademarks of an organization" (1992a:186).

Roughly, here is how it works. If an agent's "combat condition" matches (or is a prefix of) the "offense tag" of another agent, the two will be brought into combat, and the result of this combat will be determined by a "payoff-matrix" built into the particular "world" they inhabit. The winner will acquire all the letters making up the genome of the loser, storing these for later use (in reproduction, in trade, or in paying taxes). The loser is deleted from the population. The code that tests for combat looks like that in figure 20. Trading occurs when an agent's "trading condition" matches (or is a prefix of) the "offense tag" of another agent (and vice versa); when this happens, there is a transfer of resources between reservoirs. Holland writes, "Each agent is a kind of middleman, accepting resources from other agents, modifying them in some way, and passing them on to still other agents" (1994:312). When agents have acquired enough resources, they may make copies of themselves. Alternatively they may "mate" with other agents (in pairs) using a crossover operator modeled after crossover in genetic algorithms. This mating happens only when agents' mating tags and conditions match in both directions.

As of this writing, only one Echo scenario has undergone extensive testing, and this is an ecological simulation of interactions between a set of species of ant, caterpillar, and fly. Holland harvests information about this dynamic from Bert Holldobler and Edward O. Wilson's *The Ants* (1990). In this book's account, flies lay eggs on nectar-exuding caterpillars, and ants are predators on these flies. Flies and caterpillars are therefore in a combat relation, since emerging baby flies may eat caterpillars.

Ants and flies are also in a combat relation, since ants eat flies. Ants and caterpillars mutually benefit each other and exist in a trade relation: ants consume nectar from the caterpillars (without disturbing the caterpillars themselves) and caterpillars surrounded by ants are protected from the depredations of flies.

When one uses Echo, one does not see pictures of ants, caterpillars, and flies gamboling about. Rather, an array of windows is presented, allowing the user to control parameters of the run, to edit characteristics of worlds, sites, and agents, and to view graphs of population statistics. The main window provides a scrolling record of all interactions in the world. Depending on how much information the user wants, it can report on how many generations have passed, which agents are killing which, and so on. What makes the system lively, then, is not so much its interface as the language used to describe it. In Institute presentations of Echo, people persistently spoke of agents wanting things, making decisions, being predisposed to violence, and having opinions about the world. In one talk, Holland told us that agents were hostile to one another and were each devoted to getting better at "swiping the other guy's resources." In another presentation, Jones declared, "I'm continually finding out that my agents know more than I do." Everyday language is readily mixed into technical descriptions of the system, imbuing the program with a logic not easily disentangled from the common sense of persons interpreting it. Echo is thus a good model organism for understanding some cultural commitments that animate Artificial Life systems, particularly those that portray organisms as highly competitive individuals. Holland writes, "Because Echo is a computer-based simulation, it allows no unarticulated or ambiguous assumptions. The generated behavior is a precise consequence of the assumptions implemented" (1994:317). I follow the trail of these assumptions here.

COMPETITIVE, POSSESSIVE, AND MASCULINE INDIVIDUALISM

Echo begins with a "state of nature," one that might be familiar to us from founding texts in the Western liberal political tradition by such authors as Thomas Hobbes and John Locke. In these formulations, the individual is prior to its environment, prior to its resources, prior to social relations, and necessarily in competition with others. People are seen as self-determining and equal citizens, a belief that rationalizes inequalities as "natural" and inevitable results of innate qualities and differences (Pateman 1988; Collier, Maurer, and Suárez Navaz 1995). This theory of liberal individualism is inextricably tied to the capitalist economic system: "Liberal political theory emerged with the rise of capitalism, it expressed the needs of the developing capitalist class and the liberal values of autonomy and self-fulfillment have often been linked with the right to private property" (Jaggar 1983:34). Individuals are conceived as necessarily engaged in competition, for although resembling each other in having wills, they are understood to want different things and to have interests that conflict (Collier, Maurer, and Suárez Navaz 1995). These abstract, formally equal individuals are precisely the kind modeled in Echo, and they make sense to many researchers in part because such notions of individualism are codified in the law and common sense we use to think about and experience the character of choice, freedom, and the ownership of property (see Fraser and Gordon 1992).

These notions also inhabit Echo because they inflect the evolutionary biology from which the system borrows. Neo-Darwinism is itself a genre of liberal theory, treating all individuals in a species as created equal—all abstract individuals, different only in their genetic constitution and perhaps in their strategies for getting their genes into the population. And in both liberal theory and neo-Darwinism, differences between individuals are seen logically to imply competition. Keller writes,

> Much of contemporary evolutionary theory relies on a representation of the "individual"—be it the organism or the gene—that is cast in the particular image of man we might call the "Hobbesian man": simultaneously autonomous and oppositional, connected to the world in which it finds itself not by the promise of life and growth, but primarily by the threat of death and loss—its first and foremost need being the defense of its boundaries. (1992c:115–116)

In this picture, theorists "obscure the logical distinction between autonomy and opposition. In this, they support the characterization of the biological individual as somehow 'intrinsically' competitive—as if autonomy and competition were semantically equivalent, collapsed into one by that fundamentally ambiguous concept, self-interest" (Keller 1992c:116). Here, individualism necessarily entails competition and struggle with others. Neo-Darwinists see competition as the logical outcome of the fact that individuals exist in the same environment and must struggle to possess scarce resources. There are many reasons not to believe that this follows, however. As Keller reminds us, this assumes that a resource can be defined and quantitatively assessed independently of the organism itself [and] that each organism's utilization of this resource is independent of other organisms. . . . What get left out of this representation are not only cooperative interactions, but *any* interactions between organisms that affect the individual's need and utilization of resources. (1992c:122–123)

Keller continues by suggesting that "interactions [can] effectively generate new resources, . . . increase the efficiency of resource utilization or reduce absolute requirement" (1992c:123–124). But the commitment to competition over resources is so deeply written into neo-Darwinism that it is no surprise that it surfaces in Echo, especially as the system fashions its individuals as acquiring, possessing, and protecting resources in their reservoirs.[47] The political theorist C. B. Macpherson provides a historical reference for how individuals came to be seen as naturally owning things (like property and their own bodies and labor): such ideas arose because "the relation of ownership, having become for more and more men the critically important relation determining their actual freedom and actual prospect of realizing their full potentialities, was read back into the nature of the individual" (1962:3).

Echo is structured down to the most basic programming decisions by competitionist assumptions. Ironically, when these assumptions were first built in, Echo did not immediately manufacture a struggle for resources. In one run, agents emerged that had zero-length genomes, a feature that enabled them to reproduce without resources. This went directly against an intuition that taxable consumption is required for survival. A procedure was written to prevent agents from receiving the proverbial free lunch. A comment in the code explains,

```
*Watch for agents that have
*absolutely no resources in their genome—don't
*let them self-replicate or they'll fill the world
*exponentially quickly
```

Terry Jones told me that it was not so much an economic intuition as a programming decision that prompted the writing of this comment. The problem was simply that "the program quickly consumed all the memory of the computer and halted anyway." While this technical reason is valid, it cannot be separated from the political economic conviction that resources exist apart from organisms and are necessarily scarce. Even in Echo, there is no clear reason why this should be so. In a computational universe, it is possible to create resources out of "nothing" and even to give something to someone without giving it up oneself. Information is the most obvious example.

The portrait of competition in Echo is extraordinarily gendered. The individual is selfish, aggressive, independent, utilitarian, and autonomous—in fact, so autonomous that it can even exist without others at all: "an agent located at a site that produces the resources it needs can manage reproduction without combat or trade, if it survives combat interactions with other agents" (Holland 1992a:192). All these traits are associated with a masculine individualism characteristic of capitalist societies, one that fashions individuals as most concerned with interactions in a public, political space, a space in which rational individuals come together to compete and to enter into contractual relations (see Pateman 1988). Echo agents enter into combat and trade, and when they mate it is through a kind of contractual agreement. In building the basic interactions of Echo, any values associated with nurturance, cooperation, interdependence, and care of the young or infirm are factored out. They may arise out of the competitive world, but they are always secondary. The Echo agent is cut after a masculine individual that masquerades as a universal organism. Keller writes of the biological individual copied in Echo, "Much as the atomic individual in political and economic discourse is simultaneously divested of sex and invested with the attributes of the 'universal man' (as if equality can prevail only in the absence of sexual differentiation), so too, the biological individual is undifferentiated, anonymous, and autonomous—assumed even to be capable . . . of reproducing itself" (1992a:148). Echo, like the mathematical ecology Keller is writing about, treats all organisms in a species as interchangeable.[48]

Echo agents are programmed to be "rational" in their very reproduction; they only reproduce when they have accumulated enough resources. At a predominantly white male SFI workshop on "sustainable development" one person commented on this, "So they can only have children if they can afford it. . . . [N]ot a bad idea," a remark resonant with dominant descriptions in U.S. political culture of welfare mothers as lazy and imprudent in their "choices" of when to have children. The stigmatization of women caring for children with welfare moneys is directly facilitated by the ideology of competitive individualism, which sees welfare as a "handout" to people who should be "working" (as if raising a child is not working) in job relationships into which they have contracted (see Fraser and Gordon 1992). Echo reproduction is "rational" in yet another way: it is a contract between equals; it can only happen when both agents' tags and conditions match. A comment in the `interaction.c` file in Echo explains: "Mating (in Echo) must be by mutual agreement." This protocol encodes a notion that mating is (or should be) consensual, that it has to do with attraction, and that there are reproductive choices at issue. In a way, this is necessary to sell the program, because if it were not included, if mating were a unilateral decision like combat, it would be difficult to avoid the use of the word *rape*. The fact that sexual reproduction in Echo is "economically prudent" and "consensual" illustrates a vision of individuals as calculating and contracting in their self-interest. If sexual difference were introduced, it is unlikely it would trouble this masculine model; in a cultural context in which liberal feminism has sought to enfranchise women on masculine terms, the Echo agent would remain a contractual individual.

Nowhere is the masculinity of the agent in Echo more apparent than in combat. "Combat" is not a surprising term to hear since metaphor mappings between war, games, sports, races, and competition in nature are popular in Euro-American culture. Holland himself has been a fan of competitive games and has transplanted metaphors from gaming into his evolutionary theory. One male scientist brought out the masculine edge such metaphors have when he told me about watching the results of an Echo-like simulation:

> One of my most vivid memories is from very early in the [simulation] work. I had just gotten the code working with some simple per-generation stats being printed out. For about four hours straight, [my collaborator] and I sat in the computer lab watching a string of numbers slowly scroll on the screen. We would get very excited every time one of the [agents] got further down the trail, and we kept track of the best-so-far position . . . on a map I had printed out. It was like we were rooting for our favorite sports team. Every time a new top score appeared, we were excited. Every time that new top score was lost in the next couple of generations, we were depressed.

Whenever Echo is discussed at an SFI workshop, combat is picked up happily by some men in the room, who put themselves in the place of Echo agents to understand how things are working. More than once, I witnessed men say to each other things such as "Let's say that you and I fight, and I kill you and take your resources." In an interview with one researcher, I was offered the following example of how he and I might put ourselves in the shoes of Echo agents:

> The idea was that tags were phenotypic, or at least visible traits that one agent could see in another agent. The idea of the conditions was that they were things that you really couldn't see. You might be a coward, but I don't know it, whereas I can see that you have long hair, so that your long hair might be a tag, but the fact that you're a coward might be a condition. So, ideally, if I'm deciding whether or not to fight you, I can look at your long hair and think, hmm, he's probably kind of a hippie and probably a nonviolent type, and if I attack quickly, maybe I'll surprise him, and I can kill him very easily.

One SFI staff woman reported to me that every time she walked in on a discussion of Echo and heard the word *combat,* she felt like she had entered a boys' secret club.

Scientists who have constructed images of nature as hostile and competitive have often claimed that this view faces up to tough facts, is "realistic." Keller writes,

> In short time, two tacit and complementary equations were collectively established: on the one hand, between conflict, competition, individualism, and scientific realism—that is, what life is really like—and, on the other hand, between cooperation, harmony, group selection, "benefit of the species," and a childish, romantic, and definitely unscientific desire for comfort and peace, benevolence, and security—indeed, motherliness itself. (1992a:154)

People who describe nature as nurturant are dismissed as romantics. But one could argue that the view of nature as made of self-sufficient individuals is radically wrong: interests, introspective states, and motives exist only in social contexts. The notion of a presocial and isolated organism is conceptually incoherent (see Jaggar 1983:43).

Though there is a historical and scientific heritage founding our understanding of individuals as autonomous entities, it plays out in everyday life as well. In the privileged environment of SFI, for example, some people—researchers—come and go as they please. This is in contrast to most lower-echelon women staff at SFI, who work 9:00 A.M. to 5:00 P.M. days. Because researchers are encouraged to take their privileges for granted, even to the point where these become invisible, they may feel that their actions proceed directly from their wills. This ignores how much labor is done for them, labor that *allows* them to be flexible, self-determining, and independent. To believe that individuals exist prior to the social relations in which they live, that individuals "choose" to enter into these relations, is a fiction most easily maintained by those who have the most freedom to act on their desires. Men and people at the top of racialized hierarchies have been precisely those individuals who have imagined that their self-concept should be considered generically human. When this humanity is characterized by autonomy from others and when it is extended to descriptions of organisms in general, models like Echo become possible. It is not a coincidence that many people who pass through the Institute and find models like Echo plausible are people who occupy or strive to occupy locations of privilege. These people are predominantly men and Euro-Americans, and if they are not, they are expected to downplay their divergences from these categories.

Let me spell out how some women experience pressure to become "individuals" (on the masculine model) in science. In U.S. professional worlds, women are expected to behave as through they were "no different" from men. They are expected to follow "gender-neutral" codes of behavior—meaning that they are expected not to have obligations that will take them away from work or that will force "private" concerns into the workplace. To become one of the guys, women may choose not to wear overly feminine clothes (note that male clothing defines the standard for gender-neutral clothing), or may find themselves policing their behaviors to ensure that they do not "invite" unwanted sexual attention.[49] Even as women struggle to erase their gender, men are allowed, even encouraged, to engage in stereotypical male behavior; they may make sexual jokes, treat people in a fatherly manner, and so on. Many men also have wives or girlfriends who ensure that "private" obligations remain well sequestered from work. Women in the world of science must often be "female men" (see Russ 1975).[50] As a site of scientific work in the United States, SFI is enmeshed in this reality. The "individuals" in Echo make sense to people because this is a social category from their own lives.

But there is more going on here. Much of the competitive language ensconced in Echo exists because the simulation is meant to model economic as well as ecological relations. In economic applications, agents might be firms or corporations. Echo is friendly to such applications because its abstract individualism is precisely the same sort as has historically granted companies the legal status of persons. Such economic articulations reveal additional complexities in how individuals are conceived. Agents in Echo are not entirely cut after the rational, maximizing, and calculating individual of liberal theory. They are parts of a complex adaptive system that produces nonlinearities that can destabilize any permanent equilibrium between, say, supply and demand. To adapt to this fluid landscape, agents must take advantage of nonlinearities, even redeploying them as survival strategies. Causes of complexity like competition and recombination also "allow a complex adaptive system to respond, instant by instant, to its environment, while improving its performance" (Holland 1992a:197). For agent populations to survive, they must be improvisatory and adaptive rather than maximizing and rational, harboring in their genetic makeup many possible solutions to the problems of staying fit in a world of "perpetual novelty" (Holland 1995). This requirement complicates the masculine subject of modern capitalism, taking us into territory mapped by new capitalisms that favor individuals "flexible" in their modes of accumulating resources.

TRADING ZONES BETWEEN
ECONOMICS AND ARTIFICIAL LIFE

How have Echo agents been fashioned after such individuals? Part of the makeover simply mutates a long tradition of evolutionary biological borrowing from economics, dating back to Darwin's use of the political economy of Thomas Malthus to think about scarcity. But new wrinkles in the adaptive economic agent are best understood by returning to the interdisciplinary circuits of SFI. Artificial Life research at SFI is carried out in dialogue with other fields, including economics. Echo is exemplary of the sort of system through which language and techniques are traded between Artificial Life and SFI economics.[51]

SFI's Economics Research Program has its origins in the Institute's earliest search for benefactors and its contact with Citibank/Citicorp, a company known for its credit card and banking business. At the suggestion of Citicorp CEO John Reed, SFI sponsored the workshop "International Finance as a Complex System" (1986) and, later, "Evolutionary Paths of the Global Economy" (1987), the proceedings of which were both published in *The Economy as an Evolving Complex System* (Anderson, Arrow, and Pines 1988). Citibank/Citicorp has remained a major contributor to SFI, and many other information and service industry-based companies concerned with financial management have joined as sponsors. All have been interested in SFI because they have been dissatisfied with standard economics, finding it inadequate to speak to the unpredictabilities of the contemporary economy.

The Santa Fe approach to economics sets itself up against neoclassical economics. Where neoclassical economists emphasize the perfect rationality of economic agents and assume that economies are self-regulating, SFI scientists argue that economic agents always act with imperfect knowledge, that their actions affect economic outcomes, and that economies are rarely in equilibrium. The focus is on how agents act "adaptively" in a world structured by contingency and subjective judgment. In a summary of "the SFI approach," resident program director Blake LeBaron wrote,

> Rather than reaching equilibrium, this economy is seen as being in a continuous dynamical struggle of adaptation and evolution. New goods are created which change the entire economic landscape for existing production processes. Financial markets struggle toward efficiency as price patterns, eliminated by adjusting strategies, are replaced by new patterns. (1993:1)

Brian Arthur, a Stanford-based economist who works at SFI, phrased the distinction between neoclassical and SFI economics to me this way:

> Under the standard view of economics, the economy is a gigantic machine. We are standing in front of the control panel and, like a big electric power station, there's all kinds of dials and needles pointing places. The only thing is, nobody quite knows what's connected to what, and if you could just figure out where to set these dials, everything would be just nice. Now, I don't think that that's at all useful—I don't think it's correct in any sense of the word. I think it's totally misleading. The economy is not like that. The economy is like an ecology and with an ecology you think of yourself more as a forest ranger, a gardener, a landscape architect. It's evolving organically, you can to some degree influence where it goes—you can water here and weed there, you can put up fences. . . . But basically, it's a huge ecology and one that has a multiplicity of consistent patterns.

Central for SFI retheorization of economics as an adaptive system is a suite of concepts drawn from evolutionary biology. The SFI approach is committed to a "biological view of economic systems" (LeBaron 1993:1).

As researchers seek to write simulations inspired by this "biological view," they traffic in techniques made available by Artificial Life–like systems. John Holland's work is of signal importance. In his contribution to the proceedings of the first workshop on the economy as a complex evolving system, Holland outlined the parallels, as he saw them, between economics and ecology, parallels that were an early inspiration for writing Echo. He observed, "The role of utility in economics is quite similar to the role of . . . *fitness* in evolutionary genetics" (1988:120). And: "The arena in which the economy operates is typified by many *niches* that can be exploited by particular adaptations. . . . Niches are continually created by new technologies and the very act of filling a niche provides new niches (cf., parasitism, symbiosis, competitive exclusion, etc., in ecologies)" (1988:118). Holland freely mixes the languages that describe economies and ecologies, making it difficult to know where these concepts are empirically grounded. Through the medium of the complex adaptive system, envisioned finally as a kind of computer or information-processing system, the economy becomes a kind of ecology and vice versa.

This view is operationalized in a computer model of the stock market crafted by Holland, Arthur, and others (see Holland 1995:84–87). In this artificial world, sometimes referred to as "Artificial Economic Life," a specialist program posts price, dividend, and interest rates. "Traders" are represented by strategies encoded in bit strings. All traders have access to global information about prices and trends in the market, and each trader has rules it uses to act on this information. Using the genetic algorithm, new kinds of traders are produced and ever-fitter traders emerge. In an article on this project in *Futures,* a magazine for derivatives traders and money managers, Gibbons Burke writes, "Researchers have found that nature's way of adapting living organisms to the environment—evolution—can answer difficult questions. Applied to the world of finance, it yields surprising market insights" (1993:26). There are some who think that such a model might be developed to act as a computerized trader in the real market. Indeed, one person I interviewed had consulted with a company to develop pattern recognition techniques to forecast exchange rates. Similar efforts to apply Artificial Life techniques to understanding markets are under way at the Prediction Company, a Santa Fe-based company started by J. Doyne Farmer and Norman Packard, both physicists and prominent Artificial Life theorists who were once part of the famous "Chaos collective" at the University of California, Santa Cruz, in the 1980s (see Gleick 1987). The company is concerned with advising people on how to play global markets in derivative financial instruments like futures and options, employing insights derived from the computer simulation "study of complex phenomena in physical and biological systems" (Prediction Company 1993:5). At an SFI business network talk I attended, one participant reported on a simulation designed to train CEOs in decision making about mergers and buyouts. The speaker told the audience that the simulation "portrays business in a coevolutionary setting." As this sort of talk becomes more common, the "natural" logics generating computational trading strategies may become part of the world of finance itself, reinforcing the idea that finance does indeed operate according to evolutionary logic. Peter Galison's vision of computer simulations as "trading zones" through which languages from different disciplines are coordinated comes to new life here.

The relationship of these "biological" market models to Artificial Life is complex; one might say that although inspired by Artificial Life, they offer relatively little back to the field except through programs like Echo. They might simply be artifacts through which Artificial Life techniques infiltrate the business world, through which the market metaphors in evolutionary simulations are simply made explicit. But one could also broaden the definition of Artificial Life to include these systems, since, for some people, economics has become a kind of biology.[52] If we live in an information economy, it is not surprising that biological discourse should be enlisted to describe this economy; after all, these days, biology is an information science.

SFI's economic models are crucially informed by the agendas of the corporate and high finance world. The continued funding of SFI by entities like Citibank/Citicorp creates an environment in which SFI scientists' views of the economy are likely to be informed by the interests of these companies. The view of the economy that many SFI researchers hold is also shaped by their social positions and experiences within that economy. Most participants at SFI workshops are at least middle-class in economic standing, most are middle-aged or older men, and virtually all are white. Their experiences of getting loans, building houses, making investments, and playing the stock market are fundamentally informed by the way they have been perceived and created as social subjects. Most of the economic processes they choose to model are part of a public sphere that is gendered male; there are no discussions of the many tasks involved in managing household budgets, no discussions of how wage differentials between women and men might be eliminated, no talk of how welfare and unemployment might be changed to discriminate less against women and people of color. The privileged objects of study in SFI economics—stock markets and firms—are familiar to many SFI scientists from their own experience, as are ideas about how to act strategically in these domains. Researchers' conviction that there is such a thing as an economy separate from politics, human rights, and culture is a function of their privileged position as people who really do experience some decisions as "economic," others as more "social" or "personal."[53]

Simply designating the economic positions researchers occupy, identify with, or aspire to is inadequate without saying more about the kind of economic systems they inhabit and seek to understand. According to social historians of economics (see, e.g., Mandel 1972; Harvey 1989; Jameson 1991), there has been a sea change in the way the global economy is organized. The questioning of the neoclassical paradigm results from a sense that economic reality has changed. In recent decades, manufacture and production have shifted from Fordist to post-Fordist principles. Under Fordism, production was organized around large-scale industrial production for mass markets. Workers were supported with a family wage and encouraged to participate as consumers in the economy for which they produced. Responding to inflexibilities in Fordist production (strong labor unions, costly commitments to limited product lines), post-Fordist production has been characterized by the flexible responses of capital to changes in international labor laws and markets, exchange rates, and patterns of consumption. Under regimes of "flexible specialization" or "flexible accumulation," production is done quickly, in small batches, and by laborers who can be speedily fired and hired (not coincidentally, genetic algorithms are being used to organize such post-Fordist production [see Cleveland and Smith 1989 on factory flow and Easton and Mansour 1993 on staff scheduling]). Companies switch specialities quickly to maintain a standard of profit accumulation. The labor force employed by manufacturers is increasingly global, as multinational corporations circuit the globe to take advantage of international differences in labor costs, laws, and demographics. In addition, workers and consumers are no longer the same people; they may be thousands of miles, or many social strata, apart.

These transformations have been accompanied by changes in the way money is manipulated in financial markets. Speculating on futures, options, and currency markets has become important, as has putting funds into offshore financial centers. Rapid navigation through these new economic waters is increasingly facilitated by internationally networked

computers. With such computers, businesspeople can make rapid decisions about what and when to buy and sell; they can even program computers to make decisions for them. The social historian David Harvey
characterizes post-Fordist businesses as placing "a premium on 'smart'
and innovative enterpreneurialism, aided and abetted by all of the ac-
coutrements of swift, decisive, and well-informed decision-making"
(1989:157). He also notes that "computerization and electronic communications have pressed home the significance of instantaneous international co-ordination of financial flows" (1989:161). Post-Fordism relies on technologies of rapid information transfer and communications
both to implement production and design and to support transnational
banking and investment.

These are the changes around which business folk are crafting the new
languages of adaptation. As the anthropologist Emily Martin notes, "In
order to survive in this changed environment, a wide variety of human
resource managers, consultants, and authors are advocating that American corporations must become like biological systems that successfully
survive in nature" (1994:208). What this language of adaptation erases
is the very different way capital flexibility will affect people in different
social positions. For a CEO, adaptation might mean rearranging the
capital in his or her portfolio, while, for a worker of color, it might mean
accepting reduced or eliminated benefits and nastier working conditions,
or considering relocation to another country. Women workers, often torn
between obligations at "home" and "work," are particularly vulnerable
to exploitation by flexible capital. Many women take on short-term
work and because many communities of women have only been working for multinationals for a short time, may have fewer union and labor
networks in place than men (see Fernández-Kelly 1983). As the literary
critic Gayatri Spivak writes, "Whereas Lehman Brothers, thanks to computers, 'earned about $2 million for . . . 15 minutes of work,' the entire
economic text would not be what it is if it could not write itself as a
palimpsest upon another text where a woman in Sri Lanka has to work
2,287 minutes to buy a t-shirt" (1985:171). The appeal to nature is used
to paper over such contradictions and is being newly adopted by powerful capitalists as a rationalization for their practices. The anthropologist
Bill Maurer (1995) has suggested that advanced capitalist practices of
rapid computerized financial trading and manufacturing are inculcating
in some people a new subjectivity. They are coming to see themselves as

enacting natural, adaptive logics. Maurer sees SFI economics as contributing to and symptomatic of this process. This is stunningly illustrated in a quote from a president's message in the *Bulletin of the Santa Fe Institute:* "There is something in the SFI environment that reshapes our thinking and the things we say. I was reminded of the SFI effect recently while reading an interview published in the magazine *Manhattan, Inc.,* in which John Reed, CEO of Citicorp and a sponsor of the SFI economics program, described himself as a complex, adaptive system" (Cowan 1990:2). In a public lecture I attended that was sponsored by SFI, the founder of Visa, Dee Hock, sermonized about how adaptation had been the hallmark of his career and of the success of Visa. He argued that archaic industrial forms of management were in conflict with the freedom and creativity of the human spirit and that this spirit is being reawakened by the microelectronic revolution, which through facilitating communication and rapid decision making is making possible a rediscovery of our natural creativities.

The people whose confidence is being boosted through seeing the world this way are those who are already privileged. Many businessmen who come through SFI, armed with a firm sense of themselves as self-made men, enjoy being told about complexity and Artificial Life; it boosts their egos and creates the sense that they are at the cutting edge. The fact that many Institute researchers (certainly not all) humor and behave deferentially to these men, when they would *not* do so for other laypeople who lack scientific credentials, makes clear how much naturalistic rhetoric services old hierarchies. The financial markets and capital investments that SFI researchers study are of course anything but natural; they are maintained, legitimated, and protected by government apparatuses that keep definitions of private property stable and, ultimately, enforce it though law and military force.

What does all of this mean for Artificial Life? The connections are complicated. The 1960s cohort of Artificial Life researchers cares little for the somewhat older economists at SFI, but their class positions pull them toward identification with these people, especially as some begin to take on collaborations and consulting jobs with them. This institutional link is strengthened by the way systems like Echo import images of flexibility into Artificial Life simulation. Artificial Life and evolutionary economics at SFI are in a tightening embrace, a dance that is metonymic for wider convergences between languages of economics and biology and between rhetorics of market economics and environmentalism.

EXIT

Artificial worlds are gravity wells that bring together odd pairings of science and culture. Researchers' ideas stream into Artificial Life artifacts and twist back out to inform their images of themselves, producing a kind of double helix that holds social knowledge on one strand and complementary technical instantiations on the other. As I have argued here, however, the technical and cultural are never really separate; we might think of them as two sides of the same thing, two logics locked back to back on the Möbius strip describing movement into and out of the fantasy spaces of looking-glass worlds. These worlds contain a complex latticework of belief, and act as kinds of reliquaries where the bones of these beliefs wait for reanimation in the dawn of a digital age. These silicon second natures are rich sources not just for retheorizing biological life but also for researchers' reflections on their own lives. I turn now to these lives and to the meanings scientists give them as they adopt Artificial Life into their visions of the cosmos and their place in it.

FOUR

The Buddha, the Godhead, resides quite as comfortably in the
circuits of a digital computer or the gears of a cycle transmission
as he does at the top of a mountain or in the petals of a flower.

ROBERT PIRSIG, *Zen and the Art of*
Motorcycle Maintenance

Concerning the Spiritual
in Artificial Life

IN LATE 1993, a small congregation of people from Santa Fe and Los
Alamos gathered in the pews of the St. Francis Auditorium, a performance hall built to resemble a New Mexican cathedral, to listen to a
panel discussion of an art exhibition installed at the Museum of New
Mexico in Santa Fe. The multimedia exhibition documented the little-
discussed relationship between scientists of the Manhattan Project at Los
Alamos and the people of the local indigenous community of San Ilde-
fonso Pueblo. Appropriately enough, the group set to discuss the instal-
lation included an anthropologist of science, Hugh Gusterson, who
spoke about his fieldwork among nuclear weapons scientists. Gusterson
presented tales of how these scientists used birth imagery to describe the
manufacture and detonation of warheads. During the question and comment session, a woman in the audience declared that she was profoundly
disturbed by how language used to describe things technological had
fused with language used to describe things biological. She noted that
this mixing had been on the rise since World War II and had enabled us
to speak casually of computer viruses, human cloning, bombs as babies,
and so on, all while ignoring the unsettling aspects of these recom-
binations. She concluded: "It seems to me that we're fifty years into a
religion of artificial life."

Sitting in the audience, I was struck by this extraordinary pro-
nouncement, and I pondered it long afterward. While I remain uncon-
vinced that these splicings are necessarily dangerous, I want to use this
declaration to motivate a discussion of how Artificial Life science might
indeed serve as a sort of religion for some people who do it. If portions

of this book have examined how religious imagery gets downloaded into computer simulations, this chapter fixes on how Artificial Life researchers conjure religious or spiritual messages out of the work they do. Artificial Life scientists are not unlike the modernist painter Wassily Kandinsky, who, in his 1911 book, *Concerning the Spiritual in Art*, sought to ground mystical intuitions about form and color in the sensibility of science.

It is a commonplace that science today occupies a province once reserved solely for religion. In the secular humanist world, many people turn to science for solid answers to questions of how the world works, how to endure suffering, and how to make wise life choices. One person I interviewed said in no uncertain terms that science was his religion: "I have not been religious since high school. Science plays the role of religion in my life, in the sense that when I look for ultimate answers to ultimate questions, I look to science. If science cannot provide the answer, then I am forced to live my life without it." To claim that Artificial Life serves as a religion for some practitioners is to assume that "religion" is a vessel that can be filled with the specificities of this or that belief system. But this rendering of religion is in fact very historically particular. If I say that Artificial Life has a religious glow to it, I mean that it has come to perform functions that normatively Christian Western secular culture associates with religion.

RELIGION, THE SECULAR, AND SCIENCE

We inherit from the modern social sciences a representation of society as a functioning machine or integrated organism, with parts that do service for the whole. In this picture, "religion," when not seen as deficient science, is seen as "a distinctive space of human practice and belief" (Asad 1993:27), as a universal part of human life. Religion is defined as a matter of what people believe about the ultimate questions of existence, and is distinctive in the ways it anchors belief in a supernatural being or truth. Religion is concerned with the sacred, with transcendental meanings that locate life in cosmic context.

This is precisely how religion is defined in the secular human science of anthropology. Thus Clifford Geertz writes that "religion tunes human actions to an envisaged cosmic order and projects images of cosmic order onto the plane of human experience" (1973:90). For Geertz, a religion is "(1) a system of symbols which acts to (2) establish powerful, pervasive, and long-lasting moods and motivations in men [sic] by (3) formulating conceptions of a general order of existence and (4) clothing these conceptions with such an aura of factuality that (5) the moods and motivations seem uniquely realistic" (1973:90). Religion offers reassurance that the chaos of the world is not ultimately meaningless and does so in the face of various ways this chaos manifests itself: "There are at least three points where chaos—a tumult of events which lack not just interpretations but *interpretability*—threatens to break in upon man [sic]: at the limits of his analytic capacities, at the limits of his powers of endurance, and at the limits of his moral insight" (Geertz 1973:100). Religion intervenes to offer defense against these threats. And as worshipers practice religious devotion, their subjectivities are shaped by their belief—their faith—in the order that tames these threats.

In *Genealogies of Religion,* the anthropologist Talal Asad argues that this definition of religion is specific to secularized Christian society. In premodern Christian society, religion purported to offer not just moral guidance but also a picture of the natural world and a complete social context for living. "In later centuries," however, "with the triumphant rise of modern science, modern production, and the modern state, the churches would also be clear about the need to distinguish the religious from the secular, shifting, as they did so, the weight of religion more and more onto the moods and motivations of the individual believer" (Asad 1993:39). Religious dispositions became conceptually abstracted from the fields of power that once sustained them, and religious belief came to be characterized as a private matter. Religion came to be a generic term that referred to individuals' beliefs about a supreme power and about the ethical and practical ways of living that this power mandated:

> Thus, what appears to anthropologists today to be self-evident, namely that religion is essentially a matter of symbolic meanings linked to ideas of general order (expressed through either or both rite and doctrine), that it has generic functions/features, and that it must not be confused with any of its particular historical or cultural forms, is in fact a view that has a specific Christian history. From being a concrete set of practical rules attached to specific processes of power and knowledge, religion has come to be abstracted and universalized. (Asad 1993:42)

In the view held by many Westerners, religion is simply a tool for coming to terms with ignorance, pain, and injustice. This view makes any personal philosophy a religion. Asad writes, "Geertz's treatment of religious belief, which lies at the core of his conception of religion, is a modern, privatized Christian one because and to the extent that it emphasizes the priority of belief as a state of mind rather than as constituting activity in the world" (1993:47).

As I argue that Artificial Life occupies a space that might be considered religious, I self-consciously use a Geertzian definition of religion, precisely because Artificial Life scientists often use their science to meditate on those issues that have come to be generically associated with religion in secularized Christian society. What is more, like the Christians with whom they disidentify, they frequently ground their moral beliefs in knowledge of the natural. Ken Karakotsios, reflecting on religion and science, told me,

> I pretty much completely adopted the religion of science when I was a teenager, and for a long time believed that science explained, or would some day explain, everything. But in the past few years, I've come to look at this issue a bit differently. For example, I used to think the Bible was a waste, but now I think the real problem with the Bible is that it set out to do too much. It was a text for defining moral behavior, and it was also a text for defining the way the world worked. I think the moral behavior part is still as valid as ever but unfortunately gets discredited because most of the "way the world works" part is now made obsolete by science.

"But also," Karakotsios continued,

> learning about nature has given me a new view of morality. In the race of replicating genes, there really is no morality other than "replication is sacred." Killing is just part of getting food or avoiding becoming food. Sex is just a mechanism for mixing genes. Of course, for social systems to work, there needs to be a higher-level morality applied on top, so that, for example, cooperation can be rewarded. But if you look at sociobiology, this morality is still driven by principles stemming from reproduction, which is the mechanism of replication.

The view that moral meaning might be derived from the findings of science is not new. As Karakotsios points out, this is a tradition most recently exemplified in sociobiology, a strain of natural theology I will later suggest is in some ways ancestral to Artificial Life.

Another researcher, Gerald, also saw religious issues through the lens of science:

> I think religion recognizes certain aspects of the human mind and human behavior that have been hard to account for through traditional modes of

natural selection. My work is largely designed to account for these puzzles through naturalistic processes. The ultimate goal, I guess, is to demolish religion by leaving it with nothing to explain. I think only if religion is eliminated as a *theory* of the universe, and perhaps retained only as a "cultural practice" can the world be saved from fundamentalists of various sorts.

For many I interviewed, Artificial Life provided a convincing theory of the universe, one that named information processing as the organizing logic of reality. This theory gave meaning to many people's lives and work, and in ways that often resurrected very Christian themes. Before I get to how this played out, however, I want to discuss some remarkable non-Christian imagery that surfaced again and again in my interviews and in Artificial Life texts. This is imagery associated with Zen Buddhism, a tradition that at least since the 1950s has enjoyed a good measure of popularity among atheist intellectuals in the West.

ZEN AND THE ART OF ARTIFICIAL LIFE

All of the people I interviewed were atheists, and those who had religious backgrounds had been reared in a Christian or Jewish tradition.[1] The factors that led to their movement away from these traditions and toward more agnostic, atheistic, or scientific epistemologies were various. Even as they pledged their faith to science, though, most retained—or, better, claimed to have discovered in science—a somewhat spiritual sense of the world, a sense of the world as a wondrously complex and mysterious place. It is perhaps not too astonishing that these atheist scientists often compared their scientific appreciation of the natural world to perspectives informed by "Eastern" religions. Many came of age in the midst or in the wake of 1960s Western countercultural interest in Taoism and Zen Buddhism. This interest saw Eastern mysticism primarily as an alternative to Western scientific epistemology, although there were also significant currents concerned to show how the two traditions were fundamentally compatible; the physicist Fritjof Capra, for example, in *The Tao of Physics* (1975), argued that there were parallels between quantum mechanics and traditions of Eastern mysticism. Many Artificial Life researchers cited books like Capra's as important for their thinking. Asian religions may have been appealing to some of these antireligious folks because in the popular selling of Eastern mysticism in the West, these traditions have been presented as concerned only with the purest evocations of the numinous. They have been abstracted from the earthly power structures and hierarchies within which they have existed. While the special presence of Eastern religion in the imaginations of mostly European and Euro-American scientists is not surprising, what is distinctive are the ways that many of them used the notion of "Zen" to explain how they came to view computers as worlds and computer programs as potentially or really alive.

Let me provide a dash of evidence that Zen was a popular motif. In 1989, David Goldberg penned an article entitled "Zen and the Art of Genetic Algorithms" in which he suggested that Western reductionistic views of nature might be married to Eastern holistic views to yield a more complete picture of nature (1989b:80–81). In 1994, Tom Ray wrote a piece entitled "An Evolutionary Approach to Synthetic Biology: Zen and the Art of Creating Life." He suggested that the Artificial Life researcher should approach artificial worlds with an open mind, with an attitude that "respected" the artificial medium and did not expect always to see the familiar (see Ray 1994a). The titles of Goldberg's and Ray's pieces are clear allusions to Robert Pirsig's popular proto–New Age spirituality book of 1974, *Zen and the Art of Motorcycle Maintenance,* a novel about a white American man who travels around the country on a motorcycle, learning lessons about communion with the unexpected in life, roads, and motorcycles.[2] As the author treks over the American landscape, he frames a philosophical system that fuses science, religion, and humanism. Zen was an important figure in another book that many Artificial Life people found inspiring: Hofstadter's *Gödel, Escher, Bach.* Hofstadter discussed how Zen uses paradoxical sentences to stretch thought in unexpected directions, and he compared these sentences to paradoxes of logic and to recursive structures in mathematics, molecular biology, computer science, music, and art. Zen or Eastern imagery appears not just in the practitioner's occasional literary allusion; it has also been important in the popular presentation of Artificial Life. Chris Langton is routinely referred to as the Artificial Life "guru" (an appellation more characteristic of Hinduism than Zen, to be sure), and a popular article on Langton in *Rolling Stone* features a photograph of him with his eyes closed, in an apparently meditative sitting position (Levy 1991).

Zen notions were also invoked in practice to bring computer worlds and programs to life. How did they enable this magical transformation? Through the particular understanding of Zen that researchers used. As it is usually phrased in English-language writings in the Occident, Zen is about transcending dualism, the division of the world into knowing subject and known object. Zen refuses rational and linguistically structured thought as vehicles toward the ultimate truth of reality. It relies on meditative practice to dissolve the distinction between subject and world, a distinction that has been central in objectivist science. In the Western imagination, which sees the world as divided into nature and culture, this dissolution is usually translated as enabling reunion with nature. Several people I interviewed spoke of losing the boundary between self and world when they interfaced with their computational realities. Seeing computers as worlds required a kind of meditative gaze, a gaze that saw through the gauze of the computer to the dynamics that it supported. Communion with the computer allowed researchers to see that their bodies and the computer were cut of the same fabric, were fragments of the same fundamental reality (see Hayles 1995:424). "Zen" was used by many Artificial Life scientists to refer to an experience of oneness with the computer, a oneness achieved when they had an immersed yet detached engagement with a simulation. The viewing, meditating researcher would recognize that the simulation and the world outside were both perceived through the same sensory apparatuses, so that our experience of the world might be seen as a kind of virtual reality experience, a simulation, and simulations themselves might be understood as potential worlds. One prominent researcher emphasized to me that Artificial Life could help one recognize that the observer always conditions what counts as life and that, in this way, Artificial Life was in fact "more Zen than science."[3]

VISIONS

The mystical electricity through which computer programs come to life infuses the most popular origin tales told about Artificial Life, particularly origin tales that circulate around Chris Langton and his early visions of informatic life. I alight on these stories quickly to illuminate how Artificial Life owes its genesis in part to founding events that have something of a religious, mystical sheen to them. These founding events are not documented in professional Artificial Life publications; rather, they are chronicled exclusively in popular articles and books about Artificial Life. Insofar as these texts inform the lay public about Artificial Life, and sometimes even bring new people into the Artificial Life flock, they are crucial limbs of the Artificial Life corpus. The religious cast of these anecdotes has been noted by others; Richard Doyle (1997b) and the popular science journalist Ed Regis (1990) have also commented on the aura of the mystic in these tales.

In retelling these stories, I caution that although I have interviewed Chris Langton and he has respun me versions of these yarns, I rely primarily on secondary texts, since I do not want to give the impression that these stories are about Langton "himself"; they are in such wide circulation that they have come to refer more to a myth of origin and to a widespread sense that Artificial Life might be understood through appropriate introspection.

The primal moments: Langton was working at Massachusetts General Hospital performing alternative service as a conscientious objector during the Vietnam War. One night, working late at his job as a computer programmer, he left the Game of Life running on a nearby computer. Emergent structures danced on the screen:

> So that night, says Langton, the computer was humming, the computer screen was boiling with these little critters, and he was debugging code. "One time I glanced up," he says. "There's the Game of Life cranking away on the screen. Then I glanced back down at my computer code—and at the same time, the hairs on the back of my neck stood up. I sensed the presence of someone else in the room." (Waldrop 1992:202)

But Langton was alone.

> Langton looked back at the computer screen. "I realized that it must have been the Game of Life. There was something *alive* on that screen. And at that moment, in a way I couldn't put into words at the time, I lost any distinction between the hardware and the process. I realized that at some deep level, there's really not that much difference between what could happen in a computer and what could happen in my own personal hardware—that it was really the same process that was going on up on the screen." (Waldrop 1992:202–203)

Langton sensed this again, and in a very visceral way, in the aftermath of a near-fatal hang-gliding accident in 1975. In the months he spent recovering from this bone-shattering misadventure, he had a sense of his mind as a computer rebooting, and he watched himself come to consciousness as his brain reorganized.

> It was as though his mind were a computer hit by a power surge and was now rebooting and fed a new data set. Even more fascinating to Langton

was the feeling that his synapses, in his mind's attempt to reconstruct itself, were self-organizing, much as individual ants in a colony arrange themselves in a manner conducive to perform a task. (Levy 1992b:96)

"I had this weird experience of watching my mind come back," [Langton] says. "I could see myself as this passive observer back there somewhere. And there were all these things happening in my mind that were disconnected from my consciousness. It was very reminiscent of virtual machines, or like watching the Game of Life." (Waldrop 1992:209)

What Chris Langton saw was . . . *propagating information structures!* They were proliferating through neural space, traveling down his multiple synaptic pathways, and exploding into his mind like fireworks. (Regis 1990:194)

"I had a personal experience of what growing a mind feels like," [Langton] told me. Just as he had seen life in a computer, he now had a visceral appreciation of his own life being in a machine. Surely, life must be independent of its matrix? Couldn't life in both his body and his computer be the same? (Kelly 1994:345)

Richard Doyle argues that the events following Langton's crash were foundational for Artificial Life and operated effectively because they fused scientific and religious narratives:

What crystallized A-life and allowed it to emerge as a discipline, an empirical and practical science, was a combination of three vectors, one leading from a rhetoric which equated living and nonliving systems ("propagating information"), new sources of cheap and powerful computers, and a founding event reminiscent of religious, not scientific, narratives. . . . Langton's crash . . . can be seen as an origin story of a most religious kind, a combined Icarus and resurrection myth, a new story of transcendentality told in an old form. (1997b:118–119)[4]

I would add that the religious aspect of the tale works because we understand it as tethered to an *individual* mystical experience. Langton's mind is the individuated site for the birth of Artificial Life,[5] and the power of the story resides precisely in the fact that we are not supposed to understand it as a social event. It is an individual epiphany and is authorized as such because of the privatized space in which secular Christian society sees religious experience operating. It also, of course, repeats the theme of epiphany so popular in many Western tales of scientific discovery.

The particular circumstances surrounding Langton's accident allowed him to see things anew[6]—though he certainly would not recommend this path to everyone who would splinter old assumptions. As he told me, there are various ways to divest oneself of received ideas. While I am skeptical of this possibility of complete divestiture, I would agree with Langton in the following way: I think his particular epiphany easily could have been experienced by many other folks located in the same milieu. In fact, when I asked another person from the 1960s cohort for an example of where artificial worlds might exist besides computers, he pointed to his head and said, "In the mind." For this person, mind was a kind of cyberspace, full of self-reproducing, recombining, and mutating ideas.

These visions of life as an informatic process or pattern haunting the material world concentrate motifs that many of us have come to associate with mystical experience: mystics exit their bodies and gain access to a position outside, a transcendent location from which they can see the truth of the world beneath the haze of illusion. This is, in many ways, a quite Cartesian sort of mystical experience, and we might contrast it with forms of mystical experience that intensify embodiment, forms that attune the mystic to immanent truths of the world through taste, sexual feeling, or deliberately self-inflicted pain (see Bynum 1989).

The assertion that life is propagating information is, of course, also supported by the metaphors of code that have organized molecular genetics in the past few decades as that branch of biology authorized to speak of the "essences" and "secrets" of life. But what is interesting here is that such rhetoric is buttressed by private, mystical experience of its subjective truth. And this serves as one starting point for understanding how Artificial Life researchers transubstantiate their discipline into a meaningful, religious frame for existence.

SEEKING SENSE AND SALVATION IN SCIENCE

If the story of evolution is the story of replicating information structures, and if those structures are in some sense independent of the material in which they exist, then Artificial Life researchers' minds are the breeding ground for new creatures that might one day transform into the kinds of computer programs that can spawn on their own (either in the virtual space of the computer or through the bodies of robots in terrestrial or extraterrestrial space). Seeing matters this way allows researchers to insert themselves (or more narrowly, their minds) into a grand narrative of evolution that includes Artificial Life as the next phase. Artificial Life, many researchers maintain, is inevitable; evolution has produced a brand of creatures, humans, that are now able to understand evolution's informatic logic and transport it into new media. As Langton has put it in various public appearances, "Evolution has not stopped. Far from it, we are now in its employ." Artificial Life researchers are both pawns and agents in this grand and inexorable process of evolution.

LEVY: Some people are saying that evolving artificial life is almost an inevitable part of our evolution.

LANGTON: Well, I think so. All life that we know has evolved and passed on and changed. . . . My feeling is that [life] is out of our control; we are just little cogs in a much bigger evolutionary process. We're little leaves being swept down the stream, and all we can hope to do is perturb ourselves to the right or to the left to influence the overall flavor of this evolutionary direction.

Levy 1992a:37

Another person put the same thought to me this way: "There is a zeitgeist out there. No matter what we do, Artificial Life is going to happen." Another based his vision of the future on the idea that Artificial Life is waiting to be born: within the next fifty to one hundred years, this man opined, there will be an efflorescence of new, mostly artificial life forms, engineered (initially) by humans. Life will exist as pure information in computer networks, as robots, and as genetically engineered organisms. It seemed to this man that the evolutionary process that created humans was continuing as we humans manufactured via artificial means our own evolutionary successors. To be afraid of this process, he said, was perhaps understandable, but it was also anthropocentric. There were plenty of things wrong with humans that might be improved or done away with, and he would not be sad to see something "better" emerge, though he admitted that it might take getting used to the idea that "life, instead of being generally mushy and carbon based, like fuzzy teddy bears, could be shiny and metallic." In a way, he said, he felt we humans "owed it to the evolutionary process that created us" to continue its evolutionary work. In a published interview with Steven Levy, Danny Hillis put similar thoughts this way:

I guess I'm not overly perturbed by the prospect that there might be something better than us that might replace us. Because as far as I'm concerned we've just kind of recently crawled out of the muck. We've got a lot of bugs, sort of left over history back from when we were animals. And I see no reason to believe that we're the end of the chain and I think that better than us is possible. (Quoted in Levy 1992a:39)

Still another person put it to me like this:

> If we create something alive in the computer, there's going to be a lot of an-
> gry people out there, who'll start sending letter bombs to SFI. But, if we
> don't do it, just wait twenty million years or so, and people will probably
> be gone anyway, so I think all we're really doing is accelerating the process.
> I don't think that's an argument for creating something and turning it loose,
> but I think evolution's going to happen anyway, whether or not we help it
> along. We're drastically changing the course of evolution anyway, and we
> don't know how. Maybe if we're creating artificial life, then maybe we can
> try to say something about what direction we want it to head in. And even
> if two hundred years from now, there's just a bunch of robots crawling
> around the world, well, we'll still be their great grandparents. I don't have
> children yet, though, so maybe I'll change my opinion.

In providing this grand story of evolution as unfolding toward a more
perfect future, a future in which the flesh of humanity falls away to
birth the butterfly of artificial life, Artificial Life practitioners repeat a very
millenarian, very Christian kind of salvation story—and one underwritten
by a very masculine faith in technology. In *The Religion of Technology,*
David Noble argues that technology has become an icon and vehicle
for some people's quest for transcendence over earthly life and that this
"religious" function of technology has become associated with symbols,
practices, and institutions of masculinity. He argues that the Christian
story of redemption entails a return to origins and a recovery of lost
perfection, the restoration of the "image-likeness of man to God," looking "back to a primal masculine universe and forward to the renewal of
that paradise in a masculine millennium" (1997:210). "Man" in this tale
is rational and self-sufficient, akin to Adam before Eve, Adam who was
pure and who lost his immortality because of Eve and her curiosity for
things corporeal. This man will redeem himself through extraterrestrial
trips to the stars, through the production of perfect genetically engineered
offspring, through freeing the mind from the body in the practice of
Artificial Intelligence. We might see Artificial Life as a pillar of the
masculine religion of technology that Noble names. Artificial Life offers
transcendence through engineering practices that will produce humanity's
evolutionary successors as "shiny and metallic" space-faring robots. It
offers transcendence through a perspective that sees that we can leave
behind genetic legacies that have followed us, as Hillis suggests, since
"back from when we were animals." Artificial Life might be seen as a
technological practice chasing redemption, moving away from the limi-

tations of earthly existence (see also Reynolds 1993). Like the Gnostics, an early Christian sect that considered matter evil, Artificial Life scientists become possessed by a desire to deny death through identifying with the divine.

One researcher put the longing for divinity to me rather bluntly: "If I had to declare a religion for myself, it would be basically the quest to become God. I would imagine that a god is somebody who just understands everything. I think of God as being a part of everything." This person was aware that his words might be heard as exemplary of Western scientific hubris, and added that it was a kind of Zen god he spoke about, a god that materializes in a person when one realizes that "there's no distinction between you and the universe."

While Artificial Life researchers speak playfully of being gods, their tones become more serious when they speak of evolution, the power they see as responsible for organic life and for the rise of Artificial Life. They see evolution as a creative force that has recently become more clever as it has elected a few scientists to transfer its logic to new media; insofar as these scientists act as gods, they do so as agents of evolution. In many ways, of course, evolution has become for these scientists a simple replacement or synonym for God. This easy interchangeability allows researchers to rotate in and out of being evolved products of evolution and being evolution itself.[7] Seeing into the logic of evolution, at one with its mission in an almost Zenlike way, they become divine. This Zen concept is not inconsistent with an image of gods as entities that create worlds and life. Hillis sees himself as participating in a larger process when he creates Artificial Life and envisions himself as taking after God in making intelligent computational systems:

> If I put a system inside some future Connection Machine that's the right fertilizer, and I give it the seed of human intelligence by talking to it and interacting with it and telling it what I know, and it grows and flowers into a living being, an intelligent being or something like that, then I created it only in exactly the same sense that I've created [a] flower [from planting a seed]. I've made it possible for it to exist, and I've nurtured it, but I didn't make up the rules that made it possible for such a thing to exist. I mean that's the sense in which it's mystic, I mean that's what God did. God made it possible to do that. (Quoted in Levy 1992a:41)

Ken Karakotsios told me that Artificial Life framed his thinking about cosmic order and did so in a way that left open the question of whether there was a God behind the whole thing. Karakotsios's reflections are based on a prior understanding of Artificial Life researchers as gods of their simulations:

> ALife has certainly demonstrated to me that emergent, high-level phenomena such as evolution, mind, and religious belief systems can emerge from lower-level parts without the help of a designer. But this doesn't mean that a designer wasn't present to assemble the lower-level parts in the first place. Maybe our low-level parts, things like quarks, space time, and the laws of mathematics were engineered by some entity. Maybe we are part of someone else's ALife experiment. But if this is so, then we run into the old recursion problem of where did that entity come from? Maybe if we are in a simulation, our minds or senses have somehow been limited so that we can't discern that we are in a simulation.

As Artificial Life provides a story about evolution, and offers a key place for Artificial Life researchers in it, it fulfills some of Geertz's criteria for religion: Artificial Life formulates a conception of a general order of existence and clothes this conception in an aura of factuality (in this case, a factuality generated by the science of Artificial Life). It "tunes human actions to an envisaged cosmic order and projects images of cosmic order onto the plane of human experience" (Geertz 1973:90). And it accomplishes this, as Geertz would argue, through a system of symbols.

This is a system of symbols with a variety of powerful valences. The symbols that render Artificial Life potent enough to serve as a grand vision of the cosmic history and future of humanity are not, as Geertz's definition might lead us to think, a hermetically sealed set. The cosmology of Artificial Life draws from information theory, evolutionary biology, Christian religion, and many other sources, sources that are powerfully supported by scientific institutions and by durable ideologies of American masculinity and technology worship. Following Asad, we must recognize that the symbols that animate the practices of religion are never separated from the realm of power.

Geertz has argued that a religion shapes the moods and motivations of people who participate in it. No less with Artificial Life. To begin with, Artificial Life endows the world with spiritual meaning for many practitioners. Larry Yaeger told me that his engagement with Artificial Life awoke him to the mystical flavor of the world:

> It's quite "magical," if you will, that physical processes, of the kind we comfortably associate with stones and rivers, with cooking and chemical reactions, give rise to greater and greater complexity, yielding clumps of

matter that react and adapt to their environment, including some clumps that build internal maps and models of that environment, and ultimately (so far) matter that even builds such effective maps that the point of view from which the map is generated, self-awareness, becomes an integral part of the model. This is a *very* curious phenomenon. And one marvels that it is so. There is no "reason" that anything at all exists, including the whole of our "reality." But that it does, and that it works in a way so conducive to the emergence of life, is, to me, as profound and marvelous as any religious ecstasy. . . . Instead of the animism of Native American and Australian Aboriginal cultures, I've found myself pondering the validity of a sort of "scientific animism" that admits to this continuity between what we normally think of as animate and inanimate, and thus places a great value on even the most inanimate of objects for their role as building blocks, as nurturants for the most animate of objects.

Gerald argued that Artificial Life had helped him see something spiritual in the world. He hoped it would do the same for others:

I think the main effect that ALife research will have on our concept of nature will be (1) to make us think in terms of process rather than objects or states—we'll see our current world as a very thin time-slice through the ongoing streams of evolution, development, growth, change, transformation, etc.—and (2) to make us perhaps remythologize nature as something complex and wonderful enough to feel a sense of mythical awe about. In the days of Newtonian mechanistic philosophy, nature was a tawdry and clumsy affair, just a bunch of gears and pulleys. People kept transcendental notions alive because Nature was not a rich enough concept to merit spiritual feelings. But perhaps a richer biology will allow us to transfer those spiritual emotions to a real world rather than the fictional transcendental realm of god, nation-states, Marxism, high culture, philosophy, or soap operas.

He commented on how Artificial Life had already worked some of this transformation in his own life: "ALife ideas have helped me confront aging and death, in terms of seeing how perilously we are perched in this state of 'health' and 'life,' and in seeing the continuity of life as a continuous balancing act, rather than a steady state to be taken for granted. Appreciating 'life on the edge of chaos': a more Zenlike concept of living, I suppose." Another person was comforted through his researches in Artificial Life that life was an inevitable part of the universe: "Life . . . is what you expect. There's a deep longing for the generation of order as to be expected rather than an incredible series of accidents."

Artificial Life could also be spiritually satisfying because it reframed old theological questions and made new answers available: "The soul is when you take the simple things that you understand the rules of and it has this emergent behavior that is both a consequence of the rules and also not obviously connected to it, infinitely more complicated with it. That's to me where the soul is. And I think that's a much more interesting, robust place for the soul to be than off in some little corner of science which we just haven't figured out yet" (Hillis, quoted in Levy 1992a:41).

For Geertz, no less important than shaping moods is the function religion plays in generating serious motivations for action: "A synopsis of cosmic order, a set of religious beliefs, is . . . a gloss upon the mundane world of social relationships and psychological events. It renders them graspable. But more than gloss, such beliefs are also a template. They do not merely interpret social and psychological processes in cosmic terms—in which case they would be philosophical, not religious—but they shape them" (1973:124). Certainly, Artificial Life does this; not surprisingly, many people's motivations to do Artificial Life at all are formed in part by a sense that in taking up this work, they are participating in a grander scheme, answering a calling of sorts. But Artificial Life can also touch the way people think about actions in their most personal life. Gerald said to me, "ALife research certainly makes me more motivated to have children. It does tend to validate breeding as a sort of important activity—but it also gives one a kind of ironic detachment. You can step back and say, well, it's pretty bloody ridiculous to worry about spreading your genes, given your biological similarity to the other five billion people on earth." And also: "ALife can make for better sex and social life. It adds a dimension of raw intellectual appreciation to the more emotional aesthetics of vision and touch. You can see other people not just as homogeneous bodies but as 'complex systems' teeming with unimaginable intricacy."

The cosmic order of which Artificial Life is a part and which it also enacts is, like many religiously informed orders, a moral order. The moods and motivations enacted and produced through Artificial Life connect to encompassing moral schemes. And like the biblical religion that frequently serves as the ghost in its machine, Artificial Life anchors these moral schemes in nature. Gerald notes, "I hope that ALife fosters 'biophilia' and increases our general regard for the complexity and delicacy of human and animal and plant life. I hope it overcomes our Western religious heritage of 'human versus everything else' and repositions humans within the natural biological world. Perhaps the experience of creating and shepherding ALife creatures will make us better at preserving real creatures." Earlier in this chapter, I quoted Ken Karakotsios saying that his engagement with Artificial Life had helped him learn about nature and had given him a new view of morality, a view connected ultimately to his belief that genes have a will to propagate themselves. This perspective alerts us to similarities that some Artificial Life epistemology has to that of sociobiology, which also projects a sociomoral vision onto nature, a vision crafted with a missionary, salvific purpose.

A SOCIOBIOLOGICAL THEOLOGY AND MISSION

Part of sociobiology's project explicitly concerned morality. Indeed, the first chapter of E. O. Wilson's popular and canonical 1975 text, *Sociobiology*, was entitled "The Morality of the Gene." As the sociologist Howard Kaye (1986) has argued, Wilson hoped that a scientific account of the "selfish" character of genes (actually Dawkins's later adjective) might awaken in people a sense of their natural limits and might also prompt them to take command of human evolution by rationally working within (and against, when possible) these naturally fixed constraints. Wilson's speculations on human sociobiology have often been read as a rationalization of capitalist society, and indeed they have often served this function (see Lewontin, Rose, and Kamin 1984), but Kaye points out that Wilson saw his work as aimed at reform; he did not wish to justify hierarchical social relations as inevitable. Rather, he wanted to point out what he saw as their natural basis, with the hope that humans might be able to change them with this knowledge. The selfish purposes of genes should be counteracted through human-made moral laws that respectfully acknowledge these selfish purposes but that try to enforce adherence to the needs of individuals and society.

> The myths and meanings of traditional religions have been scientifically exposed, Wilson proclaims, but the vital needs of the individual, the society, and the culture, which religion served, remain. . . . The sociobiological analysis of human nature and culture, like the ethological, reveals the tragedy of man [*sic*] to be his inescapable burden of genetic relics from his evolutionary past now threatening him with destruction in the modern world they have created. The sociobiological analysis of religion in turn creates a moral and spiritual vacuum that desperately needs to be filled. . . . [I]t is Wilson's fervent belief that the science of sociobiology can solve these "great spiritual dilemmas." (Kaye 1986:128–129)

According to Kaye, as Wilson sees it, the problem with modern society is that cultural evolution has exaggerated many of the most dangerous outcomes of selfish gene behavior:

The analysis of culture as the sum of hypertrophied forms of genetically prescribed behavioral patterns that were "designed" by natural selection for the "world of the Ice-Age hunter-gatherer" implies that much of this ineradicable "biological substructure" is no longer adaptive to modern life and is now perhaps destructive. It suggests further that many cultural patterns (such as nationalism, ethnocentrism, and war) have diverged too far from their biological origins (kin altruism and aggression toward strangers) and their biological purposes and must be brought back into alignment. Until such time as human genetics presents us with the option of altering and artificially selecting our nature, our only recourse is to choose from among the hodgepodge of programmed emotional responses that make up human nature those that still seem desirable and those that do not, and then reshape the cultural superstructure accordingly. (1986:128–129)

Wilson hoped that sociobiology might provide a "biology of ethics" and broker a return to nature. Like a Protestant practice of individual salvation through hard work, appropriate knowledge of and struggle with our "natural" inclinations could lead to a more perfect social order. Sociobiology is a kind of Calvinism that sees humanity using the natural theology of sociobiology, the book of nature, as read by science, not as written by God, to frame action.

Artificial Life exhibits some of the religious mission of sociobiology. Karakotsios followed sociobiological wisdom when he wrote me in one email:

I also see mankind as subverting the agenda of the selfish gene. We are taking evolution to the next level: Lamarckian/cultural. So instead of improving our design every generation, we can do it every time somebody tells us a better way to do something. This is a several-millionfold speed-up. The only problem is that we can now adopt strategies for survival in a much shorter time than it takes to do a good fitness test.

Another scientist told me: "To a large extent we are what we are because of genetics, and now we've drastically changed the environment in which we're operating and we're trying to throw out things that genetically make sense for us, that are programmed in as instincts." In these discussions, biology is seen as a repository of wisdom that we might look to for instruction in the next steps of evolution. As in much sociobiology, culture is cast as a source of danger but also as a tool that can recognize its own folly and learn lessons from the biology that gave rise to it. Artificial Life's mission is to transfer the wisdom of nature into new hardware, hardware that can continue an evolutionary march toward progress, leaving human imperfection behind.

Some other Artificial Life researchers also sense a biology/culture tension, but their belief that evolution ultimately consists in the replication of information structures calms their worries that humanity has strayed too far from its sagacious genetic programs. Cultural change and the technological change it gives rise to may be faster than genetic evolution, but it is not necessarily at odds with it. One person reflected on this:

> We're getting more and more alienated from these things [nature] that created us. The distance between us and what made us is growing very fast, due to our technological tools. But from an evolutionary point of view, from a rational point of view, it all makes sense. It doesn't matter whether the process [life or evolution] is carried by carbon chemistry or by silicon or by robots or whatever it is. The same kind of dynamics can penetrate. I can easily imagine some kind of symbiosis between machines and man, in which you really have an ecology of machines. We have sort of an ecology of machines and man now—I mean, I have to drive home in a little while in my car. I couldn't live where I do without this car. And we have these computers, and we're all dependent on each other.

Like sociobiologists, some Artificial Life researchers believe that there is an intrinsic morality in the process of evolution and that in doing their research, they are promoting this morality. In their eyes, it can in fact be counted on to be moral. On this view, all should go smoothly as long as Artificial Life is produced in an "evolutionary" spirit—a project that is, of course, the very mission of the discipline. A key researcher had this to say in a published interview:

> I have a fairly strong feeling that the process of evolution carries with it an intrinsic fairness to all the entities that have participated. And as long as what's happening is an integral part of the evolutionary process—what's already going on—I think that that fairness will be part of the process. And when I say that the only thing I'm afraid of is this ineptness [an ineptness that humans might introduce into their Artificial Life practice], that we might introduce an element of cruelty, what I mean by this is that we might not take this next step in a truly evolutionary way. (Packard, quoted in Levy 1992a:44)

How humans would be able to take an evolutionary step in a nonevolutionary way is hard to fathom, given the argument that Artificial Life is a logical step in evolution. The argument here seems rather to be that nature is inherently moral. The view of evolution as ultimately fair offers an Artificial Life answer to the problem of suffering; in this view, suffering is part of a grander picture. This is an old theodicy—a defense of God's goodness in view of the persistence of evil—in new hardware. It is also, of course, a theodicy that is only satisfying to persons in privileged positions. And it is primarily from these positions that the missionary promise of Artificial Life—the promise to transcend the messy aspects of human life—is visible.

ESCHATOLOGY

While many people I interviewed saw nature as a guide for wise moral decision making, some had learned a negative moral lesson from Artificial Life, one that did not "affirm" anything and so might, according to Geertz's definition, not count as a religious view at all. As one Santa Fe programmer told me,

> Working at SFI has made me aware that humanity has been around a very short time. Somehow it's soothed the notion of the human race becoming extinct. It's made me feel a little more indifferent about it, and in a sense that's given me some more leeway to just live, instead of feel so worried. I view it a little more coldly than I did. We are just a species. Nothing special about us. One day I realized that I felt much more indifferent about the human race than I ever had before and I think it was just talking about species coming and going, species coming and going. That's the way of the world. That's the way of life.

Some Artificial Life researchers have looked into the eyes of this theory of things, this idea that evolution ultimately "doesn't care" about individuals or species, and they have been afraid. They see themselves as bringing about a stage of evolution that will produce creatures that may outcompete and outconnive humanity in contests for evolutionary domination. And they feel uneasy: "What I sometimes feel bad about is that I don't think that mankind is grown up enough to take this responsibility. I'm not convinced that the day we are able to initiate processes which will create life with certain properties [will initiate] a better life—I mean the good life for humans. . . . I feel in some way that I am committing sin by the things I am doing" (Rasmussen, quoted in Levy 1992a:37–38). The language of sin only highlights the "religious" aspects of Artificial Life. At the dawn of the new creation, there is the original sin of the Artificial Life researchers. Adam and Eve ate of the Tree of Knowledge, the first tree in the Garden of Eden, but Artificial Life researchers will eat of the second tree, the tree that Adam and Eve were unable to get to, the tree that will make them as gods or as devils: the Tree of Life.

In their story of how evolution has hijacked humans' working energies to engineer the next stage of evolution, Artificial Life researchers only occasionally notice that this narrative positions them as a new elite. When they speak of humanity, they are speaking of a small fraction of humanity, and they are explicitly locating themselves as the vanguard force of evolution. Artificial Life researchers operate within a frame that grants special meaning to their own action in the cosmic order; anyone not participating in the project simply becomes a bit player in their grand saga. There is, of course, another religious motif here: that of a chosen people, a special group to whom the revelations of a cosmic force (in this case "evolution") are made manifest. Artificial Life researchers become reincarnations of the ancient people of Israel or of the Mormons. One researcher made these themes explicit when he wrote me over the Internet about his fears for the future:

> I feel hypocritical. On the one hand, I feel as though I'm involved in creating a new (scientists') religion, in the sense of an energizing ideology, i.e. let's build artilects (artificial intellects) and transcend human limitations, but on the other hand, I am pessimistic for the future, because I foresee bitter ideological wars that I am helping to create. The potential intelligence of computers is vastly superior to ours, so I believe it is only a question of time before the dominant global political issue of the 21st century becomes "Who or what should become the dominant species?" I see humanity splitting two ways, those insisting that humans remain dominant AT ALL COSTS, even if it is necessary to exterminate the second group, who feel that humans have a duty to create the next higher form of evolution. There is a whole universe out there waiting for post-human exploration. Rivalry between these two groups will be bitter, because the stakes are so high, namely, the destiny of the human species. Eventually, I see the second group being banished from the planet to pursue their experiments at making artilects elsewhere. There is plenty of space in space. ALife may lead to our salvation or our doom.

In this astounding science fictional story, Artificial Life researchers are a persecuted people, but a people who have a mission to populate and colonize new lands with their offspring. This is a millenarian story, in which things get worse before they get better. As Haraway observes about such tales, "Oddly, belief in advancing disaster is actually part of a trust in salvation, whether deliverance is expected by sacred or profane revelations, through revolution, dramatic scientific breakthroughs, or religious rapture" (1997:41). All these items are delivered in this re-searcher's story; there is even a rapture for Artificial Life researchers as they are whisked away from planet Earth into the beyond.[8]

If this cowboy story of persecution and colonization is highly mascu-line, it is also shot through with highly racial—even eugenicist—tropes. The racial themes in the Artificial Life vision of the future were made manifest by the man who thought that we owed evolution a debt for cre-ating us. He said that as humans and artificial organisms begin to live together, there will likely be a lot of tension, a tension that he said would be akin to the "racial tension" that happens when "different groups jockey for position for access to resources." He told me too that there will be humans who will resist the evolution of superior machine forms of life, and these humans will be guilty of an atavistic anthropocentrism. What is interesting is that those participating in Artificial Life, for the most part white middle-class heterosexual men, get marked as the truly enlightened ones, the ones who are willing to cede the evolutionary throne, the ones who are most "objective" about their evolutionary con-dition. Those who for various structural and cultural reasons do not par-ticipate in Artificial Life are the unspoken referent when researchers speak of the people who will reject the improved life produced by Arti-ficial Life. The appeal to antianthropocentrism is made by people in po-sitions of relative privilege; only people who can pretend to be objective, who can comfortably and without a sense of absurdity announce that evolution is "fair," can imagine themselves identical to a disinterested God. Artificial Life becomes a church built of tools taken from the cul-ture of white American Judeo-Christian atheist Zen (dis)belief.

The missionary zeal of many Artificial Life scientists grows from their personal faiths in science as a path toward understanding and ordering the natural and social world as well as the promised lands of the supranat-ural and the artificial. In this way, Artificial Life is similar to aspects of New Age practice, in which alternative religious stories are blended with scientific ones to yield new spiritual libations. This is a similarity that many Artificial Life researchers would reject categorically. In my next chapter, I begin with a story that illustrates how this rejection happens in everyday life. This story leads to a discussion of the diverse definitions of life that circulate in the transnational arena of Artificial Life and to reflections and refractions on new and mutant mixings of "nature," "cul-ture," and "life" in late-twentieth-century science and culture.

FIVE

What are the characteristics of living things? At school, in biology I was told the following: Excretion, growth, irritability, locomotion, nutrition, reproduction, and respiration. This does not seem like a very lively list to me. If that's all there is to being a living thing I may as well be dead. What of that other characteristic prevalent in human living things, the longing to be loved? No, it doesn't come under the heading Reproduction. I have no desire to reproduce but I still seek out love. Reproduction. . . . Is that what I want? The model family, two plus two in an easy home assembly kit. I don't want a model, I want the full-scale original. I don't want to reproduce, I want to make something entirely new.

JEANETTE WINTERSON, *Written on the Body*

Artificial Life in a Worldwide Web

ONE FALL AFTERNOON AT SFI, I attended a lecture given by an independent researcher, Elpida Poulos,* on understanding the planet Earth as a living entity. Poulos, who was trained in physiology and interested in ecology, had been out of academia for many years and had contacted SFI after reading a popular book on complexity. Her talk, attended by SFI researchers and some people from the local New Age/alternative science community, turned out to be quite controversial and ended with a revealing confrontation between SFI scientists and the New Age contingent. I reconstruct this event to report on how SFI scientists police the boundary between science and New Age knowledge but primarily to motivate a discussion of diverse definitions of "life" in Artificial Life and to reflect on how these are inflected by researchers' gendered and sexualized subjectivities, disciplinary training, philosophical and political commitments, technological practices, and institutional locations in the transnational Artificial Life community.

Poulos, who was introduced (following her instructions) as a "recovering scientist," began her talk with a biographical tale. After working in science for some thirty years, she recounted, she became dissatisfied with the regnant scientific view of reality; its objectivist attitude bleached the world of enchantment and encouraged a denial of knowledge making as a human activity shaped by and responsible to the global ecosystem. Poulos left science and moved to a small Mediterranean island,

*Not her real name.

where she hoped she would write fiction and rediscover the communion
with nature she had enjoyed as a child. As it turned out, she could not
keep away from science and soon began searching for scientific approaches to ecology that recognized the observer's implication in observed systems. She became fascinated by the Gaia hypothesis, a formulation forwarded by the British atmospheric chemist James Lovelock
which maintains that Earth can be thought of as a living thing, as an entity that continually produces the conditions of its own maintenance (see
Lovelock 1988).[1] Some time later, Poulos discovered the heterodox writings of the biologists Humberto Maturana and Francisco Varela. Maturana and Varela are Latin American biologists responsible for framing
the theory of biological autonomy, often glossed as *autopoiesis*, a neologism that means "self-producing" in Greek (see Varela, Maturana, and
Uribe 1974; Maturana and Varela 1980; Varela 1979). The autopoietic
view holds that living systems are best understood as entities that call
forth the conditions of their own existence.

> An autopoietic system is organized (defined as a unity) as a network of
> processes of production (transformation and destruction) of components
> that produces the components that: (1) through their interactions and
> transformations continuously regenerate and realize the network of
> processes (relations) that produced them; and (2) constitute it (the machine)
> as a concrete unity in the space in which they exist by specifying the topo-
> logical domain of its realization as such a network. (Varela 1979:13)

Poulos was impressed by autopoiesis. Here, she declared, was finally
a satisfying definition of life: life as that which creates itself. Defining life
as a process of continual self-maintenance promised to include a wide
range of processes in life's cycles of origin, existence, and demise. Life
could be seen to arise from and return to inert materials. Biota and abiota
would become different stages of the same stuff, the distinction between
life and nonlife, organism and environment, would be blurred, and the
Earth as a whole might be reconceptualized as a living entity. Autopoiesis
usefully complicated the Gaia hypothesis, replacing Gaia's vestigial commitments to objectivist accounts of global balance with attention to
interactions between components of the Earth's ecosystem, interactions
that included activities of human observation.

As Poulos continued, a few scientists became visibly uncomfortable.
Mention of Gaia had gotten them to squirm nervously in their seats.
Within mainstream science, the Gaia hypothesis is viewed suspiciously,
as having little empirical basis, as being too romantic about the balance

of nature. The fact that "Gaia" is the name of an ancient Greek Earth goddess and that factions of the New Age have embraced Gaia also makes orthodox biologists leery. And Poulos's invocation of autopoiesis did little to ease her audience's discomfort; Maturana and Varela are considered to be mystics by computationalists at SFI. This became clear during the question and comment session, when one scientist demanded a mathematical formalization of autopoiesis. Following up on this call for precision, a European-trained SFI scientist, more sympathetic to autopoiesis, accused Poulos of generalizing the theory to varieties of dynamical systems that were clearly complex but not obviously adaptive or living. Taken aback by the hostility of her audience, Poulos tried to steer the talk toward the broader philosophical issues at stake. She asked why we were so interested in the distinction between life and nonlife, remarking, "We're the ones putting the boundaries around these things."

At this juncture, some of the New Age folks spoke up. (There were about six of them to the nine or ten SFI scientists. Half of the New Agers were women, and most of the scientists were men. All, like the scientists, were white.) Struck by how the scientists did not seem to be hearing Poulos, one man remarked that he had heard a similar dialogue years ago, but between Hopi elders and a UN delegation. He spoke of how the Hopi people had for a long time had a notion of the world as a living entity. The Gaia idea was not new, he said: "Indigenous peoples have had it for millennia." He concluded by saying that humans have a choice between a picture of nature as a resource for control, patent, and copyright or as something with which to live in harmony. It was an impassioned diatribe, and when he was through, another man began to clap approvingly but quickly stopped, seeing that the scientists were sitting stone still in contemptuous astonishment. One older SFI scientist broke the silence, saying that the links being imputed between Western ways of knowing and the history of industrial, exploitive culture were specious. Tempering his anger with measured speech, he said he felt the discussion was going in an inappropriate direction, touching on political issues irrelevant to science. Another SFI person added that it was misguided to mix social and political goals with the objective of science, which was "to understand."

Taking hold of the negative energy and using a kind of intellectual T'ai Chi to transform it into fuel for discussion, Poulos intervened and argued that if science is about producing knowledge useful for prediction and explanation, that usefulness is dictated by our always social purposes. She tried to establish a connection with the scientists by commenting that knowledge should be used for survival and that her own motivation for rethinking science and working for sustainability was founded on her instinctive will to survive as an organism. She expressed astonishment that the discussion was getting so heated, since she had thought SFI was interested in blurring disciplinary boundaries. She asked the audience why there should be only one science. There are many religions, perhaps all paths to the same higher truths. Why not different sciences? The talk ended far after schedule, and as scientists left, they expressed disquiet that an event like this had unfolded at the Institute. One commented that the discussants had said "nothing that's not said everyday in Santa Fe, just not at the Santa Fe Institute."

What had triggered such a hostile reception to Poulos? In many ways, Poulos's vision of "life" was not so different from many Artificial Life scientists'. Like Chris Langton (who was not in attendance), Poulos argued that life was a property that could haunt organic and inorganic matter. Like Langton, she saw life as a question of organization. And like most Artificial Life scientists, she saw life in entities not previously considered alive. The hostility came from how Poulos narrated her story, from her choice of allies, and from her contention that science should be guided by human values. It came too from using a theory associated with the New Age, an association only reinforced by the outsiders' interventions. The animosity was also conditioned by the way Poulos framed her theories as motivated by her intuitions about nature; many in the audience heard her story as one of "feminine" intuition, especially when she invoked Gaia. Unlike Langton's vision of Artificial Life, which asks the scientist to take a transcendent position over alternative biologies, Poulos's story was about forging a theory of life from a position of immanence, from inside the dynamics under description. While these different attitudes have nothing essentially gendered about them, the transcendent disembodied view is historically associated with masculinity and the immanent embodied view with femininity. Poulos's position as an older woman out of academia—and as a woman who confessed that she had been discouraged from doing science because "science was for boys"— added to the mostly male audience's perception of the autopoietic tale as insufficiently rigorous.

In the worldwide web of Artificial Life, differences over the definition of life are common. There are diverse, contested, and unstable definitions circulating in the field. If most of this book has been weighted toward discussion of the shared symbols that render the simulated worlds of Artificial Life intelligible, this final segment examines differences in the field, differences that emerge at intersections between scientists who operate with different approaches, in different institutional contexts, and with different epistemologies. These differences fracture the field, but also, in their mutual tension, hold it together.

My first task here will be to review mainstream, SFI-associated, attempts to define life as well as to record the words of SFI-allied people who found these definitions unconvincing. My next task will take me out of SFI to examine the transnational arena within which Artificial Life operates. A global network of research sites, mostly in the United States and Europe (though also, recently, in Japan), knit together the field. This network is not only increasingly realized through the package of Internet services known as the World Wide Web, but also shares a certain logic with it. Like the Web, the community is dispersed and held together with links that connect often surprising assortments of people and ideas. These concatenations are enmeshed in fields of institutional, disciplinary, and international power, so that some sites—chiefly those in the United States—have more say than others about what will count as "life." This chapter double-clicks on links to relatively less powerful sites. Most centrally, I discuss a definition of life—autopoiesis—that has been important in European venues. I am interested in the possibilities of this alternative vision and concerned to demonstrate how definitions of life move out of their local contexts only as scientists forge relationships with machines and people that accumulate cultural and scientific influence. "Life" emerges as a political as well as a biological category.

INFORMATIC DEFINITIONS OF LIFE

The definition of life occupies a peculiar position in Artificial Life, at times central, at times peripheral. Some researchers say we will only be able to judge candidate creations as living or nonliving if we have a stable definition. Others say that to seek to answer the definitional question is to chase an ever-receding horizon and that work can well proceed without a definition. Ken Karakotsios articulated this position exactly:

> I think it's a fun game to try to define life in words, but a hopeless task to do it with any scientific rigor. Life is a verb, not a noun. It's a collection of processes; a set in which some things, but not all, need to be present. I think that we can study many very fundamental properties of life without

a concrete definition of Life. As a defense to this point of view, look at all the research that has gone on in psychology without an airtight definition of "mind." We can certainly treat systems as black boxes, and observe their behavior without having a perfect definition of what's going on in the box.

Many I interviewed said that life would be something they would recognize when they saw it, a view that makes the question one of socially located aesthetics (see Lestel, Bec, and Lemoigne 1993). In spite of this relaxed attitude toward definition, however, most people also had a strong conviction that a definition of life, once framed, would need to be broad enough to include noncarbon entities. One computer scientist had this to say: "It is not important to focus on a formal definition of 'life.' But it is important to understand and agree that the word *life* refers to organization and process, not to substance and substrate. In this respect it is like the word *wave,* which is a description of the dynamics of a process, not of the medium in which it propagates."

Chris Langton has written mission statements along these lines: "The dynamic processes that constitute life—in whatever material bases they might occur—must share certain universal features—features that will allow us to recognize life by its dynamic *form* alone, without reference to its *matter*" (1989:2). And: "The most salient feature that distinguishes living organisms is that behavior is clearly based on a complex dynamics of information. In living systems, information processing has somehow gained the upper hand over the dynamics of energy that dominates the behavior of most non-living systems" (Langton 1992:42). These contentions outline the territory within which many SFI Artificial Life researchers have sought to locate their thinking about life. In "Artificial Life: The Coming Evolution," J. Doyne Farmer and Alletta d'A. Belin provide a list of properties they think characterize the living and open their catalog with appropriate caveats:

> There seems to be no single property that characterizes life. Any property that we assign to life is either too broad, so that it characterizes many non-living systems as well, or too specific, so that we can find counter-examples that we intuitively feel to be alive, but that do not satisfy it. Albeit incomplete and imprecise, the following is a list of properties that we associate with life:
>
> · *Life is a pattern in spacetime,* rather than a specific material object. For example, most of our cells are replaced many times during our lifetime. It is the pattern and set of relationships that are important, rather than the specific identity of the atoms.

- *Self-reproduction,* if not in the organism itself, at least in some related organisms. (Mules are alive, but cannot reproduce.)
- *Information storage of a self-representation.* For example, contemporary natural organisms store a description of themselves in DNA molecules, which is interpreted in the context of the protein/RNA machinery.
- A *metabolism* which converts matter and energy from the environment into the pattern and activities of the organism. Note that some organisms, such as viruses, do not have a metabolism of their own, but make use of the metabolisms of other organisms.
- *Functional interactions with the environment.* A living organism can respond to or anticipate changes in its environment. Organisms create and control their own local (internal) environments.
- *Interdependence of parts.* The components of living systems depend on one another to preserve the identity of the organism. One manifestation of this is the ability to die. If we break a rock in two, we are left with two smaller rocks; if we break an organism in two, we often kill it.
- *Stability under perturbations* and insensitivity to small changes, allowing the organism to preserve its form and continue to function in a noisy environment.
- *The ability to evolve.* This is not a property of an individual organism, but rather of its lineage. Indeed, the possession of a lineage is an important feature of living systems.

1992:818; see also Rasmussen 1992

These criteria have been used by various researchers to think about whether their systems harbor life (see, e.g., Spafford 1994; Yaeger 1994). Not surprisingly, people have found the list friendly to a positive answer. One scientist found the most abstract processes associated with life compelling: "My 'intuitive' definition of life is centered on self-replication and evolution. If it reproduces as a side effect of its own natural behavior, does so with variation, and is a member of a population that undergoes evolution by natural selection, it is alive in my book. This definition is not bulletproof, but it is a good working one for me."

Everyone agreed that attempts to fence in "life" with one definitive list were in some sense hopeless. Gerald agreed with this general mood but also saw the fuzziness of definitions of life as explicable with reference to human evolution. If Artificial Life organisms did not count as alive, this was because we had not evolved the intuitions for dealing with them:

Everybody "knows" what life is, intuitively and preverbally, and it's more important to do the research now than have semantic debates. But in the long run, I think ALife can help clarify the concept for biology, by distinguishing the critical features of biological systems (in my view: replication and adaptation) from the historical contingencies of life on earth (e.g., the particular structure of DNA, or the design of the eukaryotic cell, or whatever).

I have my own combination of intuitive ideas and scientific ideas, the latter arising from my study of evolutionary biology. So I see replication and adaptation as the fundamental almost definitional concepts in "life," but my criteria for *recognizing* life might be more pragmatic and intuitive, e.g., goal-directed movement, flexibility of means to achieve ends, use of sensory systems and movement systems, energy-processing metabolism, growth, sex, social life.

I believe humans have a universal "intuitive biology" that informs our intuitions and theories about life. These intuitions evolved to have a certain form because they were useful in that form to our hunter-gatherer hominid ancestors. These intuitions are probably not coherent, consistent or infallible, nor are they necessarily easy to generalize to systems that we did not evolve to deal with. Much of the confusion over ALife results from trying to apply intuitive biology theories to domains with which humans have very limited evolutionary experience. We have deep intuitive conceptions of intentionality, animacy, agency, motivation, and cognition, which apply very well (i.e., very usefully) to animals and people but very uneasily to plants, mechanical devices, robots, computers, and artificial systems.

Although there is a fair amount of hedging in these attempts to define life, they all allow and invite a view of life as an abstract, informatic process. One young computer scientist at SFI said to me, "I think it's good to try [to define life], but arguments about what life is will never end. Having these lists around will help people make a case that our computer software is alive, but I think that there's going to be a lot of prejudice, and a lot of people will say that if it's in the computer, it can't possibly be alive." The programmer behind PolyWorld, Larry Yaeger, also found the definitional problem daunting:

> Unfortunately, I simply do not know for a certainty what constitutes "life" (nor do I think anyone else does). If I did know for a fact precisely what defined a living process, then I could probably tell you whether the "organisms" in PolyWorld are alive or not. . . . I begin to believe that there might actually be an information-based measure of the quantitative degree of life. If this theory were developed enough to have some confidence (and more than an arm-waving amount of detail) in it, it might actually be possible to assess quantitatively just how alive the organisms of PolyWorld are (or are not). I think we all might be surprised at the answer. I even think the answer might be: more alive than a virus, less alive than a dog; I cannot even speculate on where they might stand relative to a bacterium or an Aplysia.

It should not be surprising that this informational concept of life has something to do with the training many people have in computer science. Ken Karakotsios made this explicit:

> I started out looking at ALife as a computer architect and programmer. I've been trained to look at things as processes, where the same process can be run in different ways on different hardware architectures. Also, the time I've spent exploring CA [cellular automata] has given me an intuitive feel for the emergence of higher-level phenomena from simple, low-level, interacting parts. So I have great confidence that many of the processes mentioned in my previous definition of life emerge from the interaction of simple, low-level parts.

At the same time, of course, one does not have to be a computer scientist to see things this way. Tom Ray, trained in evolutionary biology, is a case in point.

These definitions of life are vague enough to include a medley of processes—and not only computational ones—that are not now considered to be living. In fact, Langton has been enthusiastic about designating "economies, socio-political institutions, and other cultural structures" as possible candidates for artificial life, and he suggests that cultural and genetic inheritance might occur through structurally analogous processes (1989:xxii). Mark Bedau (along with Norman Packard) has proposed that life be seen as a property exhibited by evolving systems in which there is continuous incorporation of informatically based innovation. Bedau and Packard write, "The field of artificial life is searching for a definition of life; even better would be a criterion of life—a public, empirical, repeatable, quantifiable test for whether a system (possibly artificial) is alive" (1992:458), and they "define evolutionary activity as the rate at which useful genetic innovations are absorbed into the population" (1992:431). For Bedau and Packard, evolutionary activity is equal to "vitality," and therefore to "life." Organisms located in evolving populations are alive/vital, but so are systems of ideas reproduced in, say, a scientific community (see Bedau 1996). Grand assertions like Langton's and Bedau and Packard's—that virtually everything that can be described as having informatic inheritance is a living system—bunch together scientific, commonsensical, and colloquial notions of what counts as life. In fact, under this definition, even "culture"—where culture is seen (in the tradition of midcentury U.S. anthropology) as a system of ideas floating above a material substrate—counts as artificial life.[2]

Intriguingly, the expansive definitions of life allowed by Santa Fe Artificial Life—a variety incorporating computationalism, information theory, and self-organization theory—are not too different from those Poulos outlined in her talk. As in Poulos's theory, many things not previously considered living enter the domain of biology. Disciplinary and institutional power of various kinds support the informatic, computational definition and exclude theories and phrasings like Poulos's.

Not surprisingly, the informatic, computation-friendly, abstract definition of life purveyed by many SFI-style scientists is contested, often by people at the Institute itself. It draws criticism from organismic biologists, for example, who do not think "life" can be abstracted from its embodiments. While many think simulation a useful tool for testing theories, they do not see simulations morphing into alternative instances of life. During my interviews, a few researchers also voiced critiques from feminist and queer positions. While by no means in agreement on what biology should properly be concerned with, all these different people wanted Artificial Life to be more anchored in bodies and in histories that did not simply collapse into genealogy.

ORGANISMIC BIOLOGY, FEMINISM, AND QUEER QUERIES OF THE VITAL

Some biologists in Artificial Life, concerned with understanding the functioning and integrity of existing organisms, are unhappy with how informatic or abstract definitions of life often factor out bodies. One older U.S. biologist said to me that he thought it likely that life was restricted to existence in a carbon-rich, aqueous medium; he was not persuaded that form and material could be easily separated. I asked another older biologist at SFI what he thought about the informatic concept of life so popular among computational Artificial Life researchers. He said,

> It doesn't grab me, and it doesn't grab me for a very good reason. Real living entities not only coordinate information, they coordinate flows of matter and energy. They actually make things. They do actual kinds of stuff. You were a little bitty thing, and now you're all grown up. And you've gotten bigger. So you really are a real physical system out in the real world; you're not merely an algorithm that is a symbol manipulation device. You're something that's coordinating and growing matter and energy. All of the stuff part of it is left out of Artificial Life when one has the abstraction that we are looking at information processing and the capacity to process

information. So, I deeply reject that as inadequate. It won't give you a phys-ical thing in the real world. It'll just give you symbol manipulation.

He added: "The things that Tom Ray has in Tierra are not real organ-isms. One sees things in Tierra that are reminiscent of real organisms. Real life is made out of chemicals, not dancing light patterns."

Helena, also a biologist, seconded this view, though she said she was perfectly happy with the possibility that life could exist in noncarbon-based materials:

> I believe all matter has consciousness, so I have no problem with the idea that life can exist in different material substrates. But I have a hard time with anything right now that's inside of a computer. I think computers them-selves are just barely starting to get into the realm of the interesting. These little programs that are interacting inside are more like—you know, would you say that words on a page are alive just because they're information? I wouldn't. They might be able to convey the sensation of life to someone who reads them, and that's what this [Artificial Life] is doing. This is giving an impression of life. It's like, Here's a computer's impression of life. The map is not the territory, though.

Brian Goodwin, a developmental biologist affiliated with SFI and based at the Open University in the United Kingdom, has been a keen watcher of Artificial Life and an avid critic of informatic renderings of life. He has forwarded a radical theory of biological organization that offers an alternative to genocentric views of life (Goodwin 1994a, 1994b; see also Brockman 1995). Goodwin argues that organisms are expressions of possible physical forms, that evolution is a dance, not a struggle, through this space of possibility. Updating arguments first made by the Scottish zoologist D'Arcy Thompson in *On Growth and Form* (1917), Goodwin maintains that organisms look the way they do be-cause of the physical and chemical dynamics of the world they inhabit (see also Kauffman 1993, 1995; Burian and Richardson 1996). The genes that organisms harbor only tune in to already possible forms; they do not determine these forms, as neo-Darwinians hold.[3] And natural se-lection is not very important in escorting these forms across history; rather, organisms express these forms in processes of dynamic stabiliza-tion of functional and structural order. Artificial Life can provide simu-lation tools for framing a taxonomy of the range of generically possible shapes, but simulations cannot themselves be forms of life because or-ganisms are integrated wholes, wholes for which the material of which they are constituted is of utmost importance. Goodwin focuses on the organism as the fundamental unit of life, a focus that attunes him to the importance of organisms' subjective experiences of the world as they create it and it creates them.

I have argued throughout this book that the informatic definition of life, a definition that centrifuges out bodies, is enabled by a symbolically masculine collapse of life into instrumentally useful information process. Some people I interviewed also maintained this. Helena, for example, thought that Artificial Life's penchant for objectifying life came out of a tendency to represent things to that they could be controlled; she saw a connection between computers and masculinity, even as she felt that there was nothing essential about the connection: "A lot of this machine stuff is dangerous only because the intent is not well formulated. And what I mean is that it needs to come from an emotional understanding as well as an intellectual understanding, not just as a plaything. There's something about men and toys." As a biologist and as a feminist, she found distressing some of the ways Artificial Life creations were experienced as playful:

> I see various Artificial Life people kind of going off the deep end, just enjoying their playthings. It's like "oh boy, now I can do this, and then let's put these trains together and do this and attach this to that and something's broken on this train, but so what, let's put it here . . ." There's an imbalance there that comes from having [popular scientific] attention pour in on you. It's like all the ego stuff starts to reinforce the wrong attitudes.

One young woman argued that definitions of life should be informed by the value we have for living things:

> I think it's really weird to think of the things in Tierra as alive. It seems to me like there ought to be a whole lot more respect for something that you're going to call alive. That's the one thing that bothers me about Artificial Life in general is this connection to the computer. Artificial Life is so unlike real biology. Another thing I don't like is the relationship to it, of being the creator or being somehow in charge of having it happen. It's a strange relationship to have to life. Why focus interest in life in silicon? Why not pay attention to real life?

As a person with a background in field biology, she was also stunned by how math-oriented SFI characterizations of life were:

> People around here talk about biology differently than people I'm used to talking with about biology. It's much more equationized, and makes so

many analogies to physics. That makes it unintelligible to me. And that probably has something to do with my lack of background in math and physics, but it doesn't talk about the beauty of it to me. They just talk about it differently. It seems to have less life.

These two women spoke of how definitions of life should incorporate the value and beauty of life and an emotional as well as intellectual appreciation. And they argued explicitly that their opinions were informed by their experiences both as women and as biologists. At the same time, they saw nothing essential about these experiences that would make these opinions available. Nonetheless, they reminded me, objectivist views of living things have been associated with stereotypically male attitudes like rationality and impersonality. Insofar as these attitudes usually entail a forgetting of how cultural ideas fundamentally inform science, it is desirable to attend to their alternatives, alternatives that take as starting points perspectives that speak of the *experience* of being alive—an experience at once fully "natural" and "social."[4]

The masculine flavor of much computational Artificial Life is connected to who actually participates in the field. The majority are men, working in the traditionally male disciplines of mathematics, computer science, and physics—fields that have historically been concerned with "universal" definitions and that have approached their objects through objectifying them. When I asked scientists why there were not more women in Artificial Life, many argued that it was because of socialization, because of the disciplines represented in Artificial Life, and because, in a culture in which women's professional abilities are always in question, women may be wary of attaching themselves to fringe fields. One woman said, "Men, for some reason, are more willing to explore new things, they feel more comfortable just jumping into new things. Playing around. Women stick more to what they're supposed to do. Women seem a little bit more afraid to do something strange and imaginative." While she was uncertain of why this was so, she guessed it was primarily because women were not encouraged to be intellectually and professionally adventurous and were socialized to be obedient students, to fulfill the expectations of those around them (see Belenky et al. 1986). One man seconded the idea that women might have reasons to be conservative in science but said that this might be natural rather than social, suggesting that it had to do with testosterone and with "biologically based" male aggressiveness. In a curiously self-referential logic, this man thought it natural that men were in the majority in Artificial Life. Whether Artificial Life would even materialize as a discipline in a society that was *not* ordered by patriarchal power is, of course, completely unknowable. Computational brands of Artificial Life may be so implicated in masculine institutions, traditions, and epistemologies that the discipline could exist without them no better than it could without computers. If there were more women in Artificial Life as presently constituted, this would probably mean more people from cell biology and ecology, in which women are more highly represented. Whether these people would find computational creatures compelling instances of life is an open question.

The compression of organisms into information is not only serviced by masculinist disidentifications with things bodily but also enacted within a decidedly heterosexual frame, one that reduces to a recombining code all that organisms are, conflating reproduction and sex, and rendering uninteresting those aspects of life that do not have to do with reproduction. The tendency in much evolutionary theory to cram explanations of the functionality of all traits into stories about their utility in procreation reaches an apex in Artificial Life systems in which reproducing is all that programs need do to be considered alive. And when they reproduce through recombination (as in genetic algorithms), a heterosexually flavored moment becomes the only site into which any lifelike behaviors are injected. Looking at these sorts of computational Artificial Life packages from a critical lesbian, gay, or queer perspective reveals how much work goes into making such reproduction seem natural (indeed, even heterosexual, since genetic algorithm bit strings do invite interpretation as same-sexed male couples). This is work that puts to one side the role of embodiment in "life" and that mutes the ways reproduction—of any kind—is nested in always unstable and contingent relationships.

A few people I interviewed who identified themselves as lesbian, gay, or queer zeroed in on Artificial Life's fetish of informatic reproduction as indicative of the heteronormativity of the field. One woman, a biologist by training, commented that she was struck by how different Artificial Life simulations were from what she considered real life. I asked her about the divergences. She answered: "One thing is the time scale. It's sort of hard to live in the moment in these models. Basically, everything interesting happens when things reproduce, and so you have things reproducing really fast, because that's what you want to see." One man said,

I think that with more openly lesbian and gay researchers certainly you would have less emphasis on questions of why sexual reproduction evolved, with the usual implicit assumption that sexual reproduction must be more efficient. Probably there would be a lot more sensitivity to what it means for something to be alive in the individual sense rather than the evolutionary reproduction side of it. It seems like issues of metabolism would be more important. I mean, this idea that you can have something that's alive, just by virtue of the fact that it reproduces—the thing can be a black box but as long as it makes more little black boxes just like it, then it's alive—that seems terribly misguided to me. I could imagine that if there were fewer people in Artificial Life who assumed reproduction was an intrinsic aspect or goal of organisms, if you weeded out this philosophy that one of the intrinsic things to make the life of an organism complete is to reproduce, if you removed that, then you probably would have quite a bit of a paradigm shift over to questions about what it is about the structure of an organism that makes it alive. That would be the biggest shift I could see that would come out of having fewer heterosexuals in Artificial Life.

At one Artificial Life conference, I spoke with a man who called upon a critical queer epistemology to attack a simulation of sexual selection. In this model, simple genetic algorithm-based organisms had as part of their genome a set of genetically encoded mate phenotype preferences, which they could pass on to their progeny. Though the organisms were not in fact differentiated by sex, the researchers often spoke playfully of "moms" and "dads" passing mate preferences on to "daughters" and "sons." This man noted that the researchers were assuming that sexual difference could be built into the model later and that a mechanism could be crafted to guarantee that preferences only went to same-sex offspring. It may have been true that such programming would be easy, but this man suggested that the researchers were missing another possible way they might flesh out the model they had. They might use it, he offered, to show how homosexual attraction could be produced through the very same mechanisms of recombination that produced (an assumed, not explicitly modeled) heterosexual attraction. While this would allow the researchers to render homosexuality "genetic" (a problematic and conservative move in a society that seeks to naturalize unequally treated difference), it might also point them in the direction of another view of *life-as-it-could-be.*

If evolutionary stories are privileged in computational Artificial Life, this foregrounds connections between organisms that are primarily about lineage and reproductive continuity. Through a queer and feminist lens, we can see the logical problem of connecting sex always and only to reproduction, and can recognize that "reproduction" is always an unstable process, requiring relationships that can never be assumed to be natural. Of course, to mark lesbian, gay, and queer epistemologies as necessarily about sex is to flatten out other contributions they might make; queer perspectives refashion what counts as an individual, as health, as a community of orientation and connection, as economics, and as history (see Warner 1993).

The informatic definition of life among computationalists at SFI is supported by a variety of histories. Images of the genome in molecular biology, computer technology at Los Alamos, and faiths in physicalist reductionisms combine with notions of life guided by a masculine and heterosexual image of life. Definitions that emphasize embodiment do not receive serious attention. I have highlighted the place of gender and sexuality here, but might easily have spoken of other sets of Artificial Life researchers who contest along intersecting but not identical dimensions the computationalist bent of SFI. Among these are roboticists,[5] scientists who work with organic materials, and researchers who use phenomenological philosophy in their work (see, e.g., Varela, Thompson, and Rosch 1991; Wheeler 1995).

Many of these folks are associated with research centers in Europe. The European Artificial Life community is, of course, also quite diverse, and a full account of its cultural practices would take another book.[6] I contrast "Europe" with SFI—and SFI stands, for some Europeans, as iconic of a "U.S. approach"—because several people I interviewed did so, and because the contrast maps roughly onto distinct views of the relation between life and computation. In what follows, I use "American" and "European" as tools for thinking about difference but also retain a certain caution about them. I take the point of the scientist who said, "The European way of ALife is more philosophical and is closer to physics and the American way is more computational, more simulation based. But you have to go by groups and individuals. You can talk about different meetings and the people who organized them and who they invited." And the warning of the person who told me, "I don't think you can talk about an American way of doing ALife. It's more a question of your personal, educational background. Do you come here as a biologist, a computer scientist, an engineer, doing robotics? This makes a difference in perspective. Computer scientists may tend to have a more Platonic view about what a living being is, working with information structures, and they are not so interested in getting stuff done in a mechanical way." At the same time, disciplinary differences, or more reductively, individual differences, are not all there is to it. One researcher, originally based in Europe but at SFI during the time of my fieldwork, made this clear to me when he glossed "American" and "European" definitions of life and located himself as persuaded by both: "The main contending definitions of life in the artificial life community are (1) evolvability (which includes of course reproduction) and (2) 'self-sustainability' or 'autopoiesis.' I do not feel the need to choose between these extremes. My research focuses on evolution and I am therefore a bit biased towards the first definition."

I rotate my discussion around the two definitions presented here. To flesh out the character, tone, and implications of the debate between these positions, I present a vignette from a workshop I attended in the Basque country. Entitled "Artificial Life: A Bridge Towards a New Artificial Intelligence," the assembly was in many ways an attack on SFI-allied computational approaches to understanding life. This gathering of about fifty people was quite different from other Artificial Life workshops hosted in Europe (most are similar to their American counterparts, with many demonstrations of simulations, etc.) and spotlighted tendencies that make European Artificial Life distinct from its American cousin. And it gave Francisco Varela, an organizer of the workshop and a man who has been centrally and powerfully involved in each of the European line of conferences,[7] a platform to outline what he saw as distinct about the "European" program and what he saw as deficient in Artificial Life projects as articulated at SFI.

THE EUROPEAN ARENA
AND THE NOTION OF BIOLOGICAL AUTONOMY

The workshop took place in December 1993 in the stunning crescent-shored city of Donostia, the Basque country (also known as San Sebastián, Spain). In the last few years, a European school of Artificial Life has emerged which, while deriving inspiration from across the Atlantic, has its own histories and questions. Participants at the Spanish/Basque workshop pointed out that Artificial Life was not born out of the blue in New Mexico but grew from histories of cybernetics, Artificial Intelligence, self-organization theory, and theoretical biology. They argued that

Artificial Life is properly the use of simulation techniques to test intuitions about aspects of life, and will get nowhere if it is seen as the attempt to transform models of life into the thing itself.

For these scientists, computational understandings of life were as problematic as computational conceptions of intelligence. Varela contended that living systems produce the identities they have through embodied histories of "structural coupling" to an environment that they themselves help call forth. Understanding a living thing requires that the reference point of the system be taken into account, not that the system be objectified. The Hungarian theoretical biologist George Kampis, based at the University of Tübingen in Germany, argued that just because life and computation can be described as processes does not mean that life equals computation. In "Situated Embodied Autonomous Agents: A Natural Artificial Intelligence," the roboticist Tim Smithers of the University of the Basque Country (via universities in Brussels and Edinburgh) spoke of building robots that acted effectively in an unpredictable real world and did so without using internal representations of that world. The Montreal-based philosopher Evan Thompson declared that in living systems form and matter cannot be separated: "'Strong' ALife is a bizarre mathematical Platonism in which only form matters."

Much of the argument against life as a computational process took as its basis Varela's concept of autopoiesis, which he first developed with his teacher Humberto Maturana in the 1970s. Maturana and Varela hold that living systems are best seen as entities occupied with maintaining the organization of which they are an embodiment. Their autopoietic processes of self-production or self-organization unfold over history (both evolutionary and individual) and cannot be understood except through that history. Moreover, that history must be told from the point of view of the system in question. The processes that maintain an organism's identity cannot make sense except from within that identity, because that identity is fundamentally organized by its particular history of being situated in a domain of interactions. Hayles writes that autopoietic systems' actions "are determined solely by their structure, which can be understood as the form their organization takes at a given moment" (1995:415). This means that organisms create their own world and that worlds can only be defined with respect to an organism: "Reality is not cast as a given: it is perceiver dependent, not because the perceiver 'constructs' it at whim, but because what counts as a relevant world is inseparable from the structure of the perceiver" (Varela 1992:330).

This grounding of bioepistemological understanding in the organism's experience does not mean that autopoiesis cannot be formalized; indeed, Varela frequently describes autopoietic systems as machines, and has even offered a computational model of autopoiesis (for a recent version, see McMullin and Varela 1997). In *Vida artificial*, the computer scientists Julio Fernández Ostolaza and Álvaro Moreno Bergareche provide a useful elaboration (see fig. 21 for their representation of the computer model):

> The definition of the identity of the system is given by its organization and not by its structure (*organization* is the collection of relational properties of components; *structure*, the collection of material properties contingent with respect to organization). This key distinction between organization and structure brings the idea of autopoiesis into the domain of the formal. Autopoiesis does not treat those problems specifically derived from the materiality of the components. (Fernández and Moreno 1992:31; my translation)

Here is a clear connection to computationalist strands of Artificial Life research. As in Langton's view, materiality is factored out in favor of a focus on the formal. In some of his earlier writings, Varela also suggests, like the computational cohort, that there can be a universal understanding of "life": "The characterization of living systems as physical autopoietic systems must be understood as having universal value, that is, autopoiesis in the physical space must be viewed as defining living systems anywhere in the universe, however different they may otherwise be from terrestrial ones" (1979:44). Varela and other proponents of biological autonomy depart decisively from many of their computationalist friends, though, in arguing that the logic of the living is *not* a logic of information. Neither is it a logic of control; autonomous systems are not organized into levels or hierarchies. This is the explanatory artifact of an external observer. There is no part of the autopoietic system that is in control, least of all the DNA. As Varela put it at the workshop, "DNA is not a code. You can write it into a computer, even treat it as a code in certain experiments, but you haven't therefore captured an essence. And if you copy the code into a computer, this does not mean you have transferred its functioning logic." Interestingly, the autopoietic attitude holds that reproduction is not an essential property of organisms; minimal living systems need to maintain or produce themselves, and reproduction is a subset of that activity; the evolution that this gives rise to is not an essential feature of living systems.

These days Varela argues that living systems are characterized by their *autonomy*, by the fact that they maintain their own organization, and

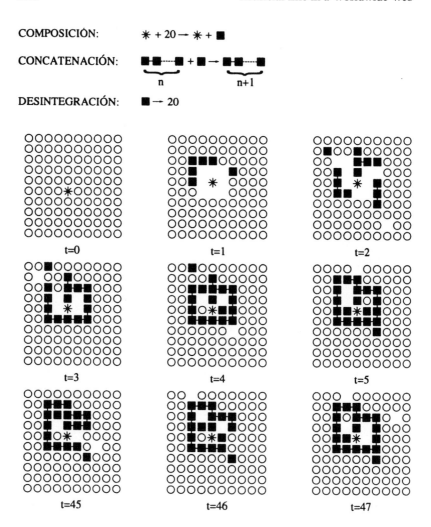

Figure 21.   "Simulación computacional de un sistema autopoiético." From Julio Fernández Ostolaza and Álvaro Moreno Bergarche, *Vida artificial* (fig. 4). Copyright 1992 Julio Fernández Ostolaza and Álvaro Moreno Bergareche. Reproduced with permission.

that *autopoiesis* is the sort of self-production proper to single cells. This clarification has not been widely heard. Most people still conflate autonomy with autopoiesis, as evidenced by Poulos's use of autopoiesis to discuss the Earth. But the claim that autopoiesis is a formal process means that most articulations do admit of abstract definitions, and there are those who would say that even noncarbon systems can be autopoietic—and, indeed, Varela is among them. As he said at the workshop, computational views of life are wrong mostly because they leave out the way living patterns result from particular histories and contingencies: "If it's silicon contingency, if it's tin can contingency, fine with me. What you cannot abstract out, centrifuge out, is that kind of process of situation that only comes from history" (see also Gould 1995).[8]

Brian Goodwin, who was not present at the workshop but who allies himself with Varela's approach, voiced a similar opinion in an interview I conducted with him:

> Artificial Life for me is the attempt to understand and to program, using mathematical and computer simulations, tests of the conceptual accuracy of the abstractions that you're using to express these properties [of living systems]. So, Artificial Life is the exploration of models that have this quality of a part making a whole, under the actions of contingencies of the environment or of heredity giving rise to a diversity of forms, which have an intrinsic relation, one to the other. So, for me, the project of Artificial Life is to describe these fundamental properties of organisms and to, in some sense, create, through simulation, the set of possible forms that constitute living beings. . . . I find [the possibility that simulations could count as real life-forms] speculative. I don't think that life is entirely captured by algorithmic relationships. There is something about the way these properties are instantiated, are embodied, that gives them particular qualities. Composition is in no sense irrelevant. The material properties are quite essential.

The notion of biological autonomy made many scientists I spoke with wary of the claim that programs were alive. One European organismic biologist told me that the fundamental distinction between programs and organisms rested on the notion of autonomy and argued that programs do not have autonomy because they need an interpreter to work, a thinking subject who continually enacts a correspondence principle between symbol manipulation and the context within which these manipulations are relevant.

Participants at the workshop were keenly aware that the naming of Artificial Life on analogy to Artificial Intelligence had encouraged computational descriptions of life. As people operating on the periphery of a U.S.-centered discipline, they also realized that the institutional power of SFI was decisive in setting the field's methodological course. A few European researchers thought American scientists could afford to remain ignorant of the diverse historical precedents for the field because of their privileged position in the global economy of science.

There are a variety of reasons Artificial Life has a different flavor in Europe. It has a more philosophical tint in part because science education in Europe includes more philosophy and because, as one European-based scientist put it to me, "It is true that in places like France, like Germany, like Italy, it is easier to engage in this sort of philosophical examination in public. In other words, if I publish a philosophical book in France, that's considered normal. If I do it in the States, you are sort of tolerated, and that's OK, but it doesn't even count on your résumé." Another person said that things were different in Europe because one did not always have to scramble for funding and demonstrate the utility of one's project: "The single most important difference between Europe and the United States is the way things are funded, because in Europe, it's been incredibly difficult to get tenured positions, but once you do, you can do whatever you want. In the United States, it's much more fluid, and you have to convince some funding agency that what you're doing is good." The focus in Europe on phenomenologically based notions of biological autonomy is also the result of Varela's involvement. Varela has been a student of phenomenology for a long time, and it tells on his science. His influence in organizing the European field is strong; based at a powerful research institute in Paris (CREA) and connected to networks in various countries, he speaks at least Spanish, French, German, and English, a fact that makes him able to net together different constituencies in Europe.

It would be easy to say that the informatic definition of life purveyed by SFI enjoys more power than the autopoietic or autonomous definition simply because it is linked with computational technology, with a materially and symbolically powerful digital metaphor for life, with the power of U.S. institutions, and with the clout of theoretical physicists linked to Los Alamos. And it is too easy to say that computational and autonomous approaches are completely distinct, since there are many points of overlap and some attempts to bring them into dialogue (see Emmeche 1997). But the fact of U.S. scientific power is inescapable. Many scientists in Europe have attached themselves to Artificial Life to gain an international audience, even as they disagree profoundly with the epistemological and political implications of computationalist views of life.

The informatic definition of life is one exponents claim will give us a transcendent view of biology. The autopoietic definition, in contrast, tends toward a more local definition of life, one that is perhaps more about immanence. Only living things can care about or attempt to define "life," and because living things are always changing, always summoning new worlds into being, no final definition of life can ever be possible. As Varela put it at the workshop, "Life, whatever we want to call 'life,' is not one property, thing, or process." At the same time, we as living entities are responsible for our definitions. One European biologist summarized this view to me in a phrase that well applies to nonhuman and human organisms: "You will live in the world that you participate in bringing into existence." For all this attention to the relations that constitute living things, however, theorists of biological autonomy still have moments when they tend toward the reification of organisms, moments when "life" remains a thing in itself, ordered by formal relations that maintain organisms as closed, self-referential entities (Ansell Pearson 1997:130; Doyle 1997a). Even so, these relations always contain pointers to an enabling "outside." As Francisco Varela, Evan Thompson, and Eleanor Rosch put the political point in their book on cognitive science, *The Embodied Mind*, "Freedom does not mean escape from the world; it means transformation of our entire way of being, our mode of embodiment, within the lived world itself" (1991:234).

At the third European Conference on Artificial Life, held in Granada, Spain, in June 1995, the Belgium-based philosopher Isabelle Stengers delivered a talk about how understandings of life should emerge from researchers' caring relationship with the objects/subjects they seek to describe (Stengers 1995). She argued that as Artificial Life scientists engaged in the production of items they hoped would surprise them with lifelike behavior, they would be pressed to acknowledge that any definition of life must implicate an interpretive interaction between knower and known, an interaction that must always call into question boundaries between what is intrinsic and extrinsic to organisms. Stengers said that this questioning—this flow of interpretation that troubled boundaries between inside and outside—should involve the "heart." She borrowed the figure of the heart from an intriguing quote by the Artificial Life scientist Stuart Kauffman, who used it in defining his own interests in being a theoretical biologist. Stengers wrote,

> I quote [Kauffman from] Steven Levy's book, *Artificial Life:* "I've always wanted the order one finds in the world not to be particular, odd, or contrived—I want it to be, in the mathematician's sense, generic. Typical. Natural. Fundamental. Inevitable. Godlike. That's it. It's God's heart, not his twiddling fingers, that I've always in some sense wanted to see." . . .
> I would propose that the God whose heart Kauffman wants to see, meaning the model for the theoretical understanding we can conceive of as demanded by life as such, is not God the once and for all creator of the world

(or microworlds) but God as It is Itself transformed through Its actual relationship with Its creation, a God in the process of becoming together with
the actual process of creation in which It participates. (1995:8–9)

Stengers argues that the metaphor of the heart demands that we think
of how our concepts of "life" are emotional, based on interactions: "It
seems to me that 'heart' in its many meanings is related to some kind of
an 'inside,' but not to a self-sufficient closed inside. It is related to the
way this inside is actually, and not potentially, interacting with the outside" (1995:8). For Stengers, life is a process defined in epistemological
partnership between scientists and living systems: "The interaction, not
the bottom-up explanation, is the primordial conceptual problem, a
problem which scientists may experimentally address only if they do not
forget about the active, meaningful part they play in the emergence of
the artificial beings which they hope will mimic life" (1995:12). Stengers
hopes that biologists who recognize this relationship will not shy away
from learning that care, trust, and love are viable tools for doing biology, for re-cognizing the living. What kind of maker will artificial life require? Stengers answers: "My answer then is, a maker with a heart, not
a sentimental heart but a caring one, because heart is involved each time
'as if situations' turn into actuality, each time imagination turns into innovation" (1995:13). Stengers has no problem with the possibility that
Artificial Life scientists may one day think of their creations as alive, but
she hopes that by the time they do, they will have engaged their emotions rather deeply, a prospect she optimistically expects will mean that
the way they practice explanation will have changed. In her talk, Stengers
carefully allied herself with Varela, a move that assured her points would
be heard at this European conference. Later in the day, I was among a
group of people who asked Stengers if practices of caring in biological
work might derive inspiration from feminist critique and activity in the
life sciences, to which she responded, "Of course."

These understandings of life depart critically from mainstream and
SFI-allied definitions and for some, attach to concerns with social responsibility and justice. The European biologist John Stewart made this
connection clear to me:

> It's very important to have a definition of life. The definition that I find at-
> tractive is that of Maturana and Varela. That life creates itself. . . . One of
> the most important consequences of the autopoietic view bears on the
> whole question of our own relationship to the world, and the whole ques-

tion of objectivism. The classical standard point of view is that there is
an objective world, reality, that is what it is, completely independently of
whatever cognition we may have about it and that therefore the task for
cognition is to be an adequate representation of that predefined reality, and
so you're going to be thinking of ways of setting up correspondence relationships and ideally an isomorphism between the representations that you
have and the real world outside. And that goes along with a fatalistic attitude. It encourages scientists to think of themselves as working to produce
representations that correspond to reality as it is. It sidesteps the whole
question of the social responsibility of scientists. Francisco's idea about
enaction or constructivism is that that is *not* the way cognition functions.
When a living organism comes into being it creates the environment in
which it exists.

This epistemology eschews objectivism but not realism. There is a real
world: the world of contingent and changing experience, the world we
produce, the world we materialize.

Brian Goodwin found autopoiesis useful for thinking holistically, and
he took this as the project of Artificial Life, a project that he linked to
his political concerns:

> For me, there's a very direct connection between the preoccupations that
> I've defined with respect to what I see as the essential properties of living
> organisms and what is called health and healing, because making wholes
> from parts, reconstituting dynamic wholes from disturbed parts is what the
> healing process can be said to be. And I think that goes directly into issues
> about the environment as well, because what we have done is to lose sight
> of the properties of integrated wholes, because we don't understand them,
> and as we get to understand emergent properties of wholes, we'll under-
> stand the stability and fragility of ecosystems. The two major crises of our
> time are health and the environment—and by health I mean not just indi-
> vidual health, but community health, and cultural health, which is an ex-
> tension of that, and which integrates, of course, with the environment. So
> the two are—or should be—I think, if you like, the crucial practical foci of
> Artificial Life: health and the environment.

Goodwin picks up similar points in *How the Leopard Changed Its Spots*.
He warns, "A biology of parts becomes a medicine of spare parts, and
organisms become aggregates of genetic and molecular bits with which
we can tinker as we please, seeing their worth entirely in terms of their
results, not in their beings. This is the path of ecological and social destruction" (Goodwin 1994b:232). He argues that a biology that puts organisms and their experiences at the center of inquiry can be different.
Here he resonates with Stengers: "A science of qualities is necessarily a

first-person science that recognizes values as shared experiences, as states of participative awareness that link us to other organisms with bonds of sympathy, mutual recognition, and respect" (Goodwin 1994b:237).

Artificial Life epistemology, viewed through the lens of biological autonomy, alerted many people I interviewed to issues of self-reflection. John Stewart said, "It's considered part of being a scientist to divorce yourself from subjectivity, and in fact anything that might give meaning to what you're doing. And it's all the more insidious because you don't have a special lecture course on being objective. It's sort of implicit, but all the more effective for that." Stewart felt that the concept of biological autonomy, with its stress on grounding explanation in experience, afforded a way of doing science that was not about scissoring subjectivity out or eliminating social accountability.

There are a variety of reasons scientists used autopoiesis to ground explicitly left political thought. Among the Europeans I interviewed, post-1968 engagement in European struggles over alternative education, the formation of autonomous communitarian living situations, meditative practice, and other issues helped them outline what they saw as a radical biology. By emphasizing the political attachments of autopoiesis, I do not mean to suggest that computational Artificial Life researchers do not share political sensibilities with the adherents of Varela. Computational Artificial Life researchers have their own tools for thinking their relation to the world, and many see their engagement with simulation as enabling participatory views of the world. The articulations that bind together people who feel that they disagree on fundamental issues are complex. Because the cultural tendencies that shape Artificial Life crosscut one another at odd angles, one can find people saying similar things with different tools, or different things with similar tools. It is this intradisciplinary diversity, supported in part by an epistemology that questions what "life" is, that prompts some to think of the field as one of the first "postmodern" sciences. This contention deserves attention, since it affords a path to thinking about how Artificial Life both shifts and solidifies the scientific ground under nature, culture, and life.

ARTIFICIAL LIFE AS MULTIPLY MODERN, POSTMODERN, AND AMODERN

Many scientists have been lured to computational Artificial Life by its use of simulation to explore theories of real and possible worlds. Several have celebrated the possibility of observing evolution quickly and in well-defined spaces in which one can tinker with initial conditions. And some have discerned something distinctively "postmodern" in this kind of experimental/simulation science. Claus Emmeche notes that Artificial Life practice can undermine an objectivist epistemology of the world that holds that reality is ultimately one thing and that there can be only one definition of life: "Artificial life must be seen as a sign of the emergence of a new set of postmodern sciences, postmodern because they have renounced or strongly downgraded the challenge of providing us with a truthful image of one real world, and instead have taken on the mission of exploring the possibilities and impossibilities of virtual worlds" (1991:161). This characterization of Artificial Life as postmodern depends for its legibility on a tacit understanding of "modernity." "Modern," of course, refers not to things "contemporary," "up to date," or "advanced" but rather to things shaped within a cultural historical project called "modernity," commonly understood to have begun sometime during the Renaissance in Western Europe. The project of modernity might be characterized as concerned with developing rational, consistent, objective, and total representations of a world assumed to be ordered according to an all-pervasive, coherent logic (see Harvey 1989). The modern quest to represent the world has operated through practices of experiment and theory in science and art and has been bound up with attempts to use such representations to predict and control the world. The technologies and epistemologies of modernity have often been tied to hopes for a future of progress, a future in which humanity is unshackled from old superstitions. In practice, of course, modernity has also led to totalitarian efforts to contain and police realities and people into categories that can be colonized, regulated, or even exterminated if deemed necessary. Modernity has multiple edges.

Artificial Life is a child of this project. Practitioners' attempts to produce a theory of life that can be true anywhere in the universe are cast in the mold of true modernism. The contention that life will be a property of the organization rather than the material constitution of entities indexes a commitment to the idea that life can be captured in formal representation. Artificial Life is also shaped by a modernist pursuit of perfection, a quest to harness the logics of nature in the aid of utopian cultural projects. And the fact that Artificial Life investigates several possibilities for how life might unfold does not necessarily wrench Artificial Life out of a modernist frame; while modernity may have been founded on a search for the one true representation, it has long been friendly to multiple perspectives on what is assumed to be one underlying, eternal reality. Stephen Jay Gould (1995) has suggested that the compacting of "life" into the gene is a sign of a modernist mood in twentieth-century life science, best exemplified in the Ur-text for informatic portraits of life, Erwin Schrödinger's 1944 *What Is Life?*

Almost without exception, researchers who tagged Artificial Life as "modern" used this designation in a derogatory sense. Brian Goodwin put his assessment of the modern aspects of Artificial Life to me this way:

> What gives me the shudders is the notion that Artificial Life is simply playing games with relationships that are suggested by living organisms so that we can replace the natural with the artificial. That for me is a Faustian nightmare, and it's the same nightmare that has dominated modernity. Manipulation. So, that's why I emphasize that a different view of the significance of Artificial Life is that it can actually transform our relationship to nature in very constructive ways.

This different view was one Goodwin linked with "postmodernism," a term that he, along with many others, almost always used in an unequivocally positive way. Goodwin continued:

> On the one hand, Artificial Life is still playing this game of modernity about paradigms. In other words, it's trying to establish a new paradigm that will supplant the other paradigms. I see that as the imperialist, authoritarian mode of science that characterized modernity. Of course, in a postmodern era, what you've got is not new paradigms but a plethora of paradigms. It's very ecumenical in that sense, a whole range of possibilities. You do pay attention to other levels, more subtle levels, that are related to quality of life, and that's what's missing from modernity as a scientific attitude, if you like. It's concerned with quantity rather than quality, and the shift from a science of quantities to a science of qualities doesn't mean you lose the science of quantities, but it is a part of a more extended science, which includes qualities. That's what the postmodern emphasis is: to be much more relaxed and much more accepting of different ways of proceeding, because they all have their legitimacy in their contexts.

As Goodwin characterized it, the postmodern view would undermine scientists' certainty that "life" and "nature" were ultimately separate from human descriptions of them. As he put it later in our interview: "We are participants in making nature what it is. The things we understand as life actually have the status of the artificial. They are artifacts of our own thinking. Artificial Life will force upon us the realization that science is our construct." Goodwin saw Artificial Life as a project of questioning foundations and used postmodernism as the optical instrument with which to fashion this view. This connects with his commitment to understanding science as always located in culture: "Science, after all, is not a culture-free activity. The point of recognizing this and the influences that act within current Darwinian theory is simply to help us stand back, take stock, and contemplate alternative ways of describing biological reality" (Goodwin 1994b:33).

Emmeche (1991, 1994) has been a vocal commentator on the postmodernity of Artificial Life. He follows a standard definition of postmodernism, one that paints it as a tendency to question the totalizing frameworks of the modernist mood, to embrace rather than eschew fragmentation and multiple worlds: "Artificial Life can be seen as a *deconstruction* of our present rational conceptions of life as a unitary phenomenon, constituted by a single universal set of 'generic' properties" (1994:557). And: "Alife research reveals that our concept of life is not a single one, as we would like to think, and that no simple set of fundamental criteria can decide the status of our models and constructs when these are already embedded in specific preconceptions of what constitutes the aliveness of natural and artificial creatures" (1994:557).

Emmeche sees that Artificial Life may actually help make some of the tensions in biology explicit:

> Artificial Life may help us see that the idea of universality of the fundamental principles of life may be a presupposition, a metaphysical prejudice with a questionable basis. Traditional biology has been haunted by a lot of conceptual dualisms and metaphysical contradictions pertaining to the methods of investigation as well as the subject matter: the dualisms between structure and process, form and function, part and whole, inheritance and environment, contingency and necessity, holism and reductionism, vitalism and mechanism, energy and information, concept and metaphor. The construction of Artificial Life can help to dissolve these dualisms, or maybe combine or reinvent them in more fruitful ways, thus giving rise to new ideas about the nature of living beings. (1994:559–560)

In an interview I conducted with Emmeche, he connected the postmodern tang of Artificial Life to recent social trends:

> Life on earth is not the only life you could have had, so to speak. But this is not only ALifers saying that. Stephen Jay Gould says the same thing in his book *Wonderful Life*. If you replay the tape of evolution, you would probably find some other creatures around. So, life could have been very different. I think that Gould's term "contingency" is very apt in this context

because contingency is also a very modern or even postmodern experience of modern man or postmodern man. We could in our own lives have done our living in a lot of other ways, and of course this can make people rather scared or make them depressed, or make them, I don't know what. . . . I think that this also connects to the status of the modern worldview, because there we have also different kinds of patches of very fragmented worldviews, so to speak. We do not have one coherent view. We have astrophysicists telling us about the whole universe and the big bang. We have different biologists telling us other kinds of stories. Of course, in a very general sense, you can put all these together and see that they do not contradict each other. But, my point is that as ideas they do not fit into each other very well, or very easily.

An older U.S. biologist I spoke with took a postmodern emphasis to be indexed by the smudging of boundaries between disciplines in Artificial Life:

> The step that's occurred in the formation of a field called Artificial Life is that the boundary between what is science, what is experimental mathematics, what is interesting games, and what's an art form has gotten to be very unclear. That doesn't mean that it's not all interesting. And it doesn't mean there's not good science being done. I think there is good science. I just think it's a little bit less clear what the science is.

Pronouncements like these eventually refer to simulation as the engine endowing Artificial Life with its postmodern spin. The simulacrum—the copy without original—becomes valued as a thing in itself, as a possible instance of life itself, and the ultimate ontology of life starts to shimmer like a mirage. Simulations slither away from ephemeral natural referents and become interpreted as new instances of *life-as-it-could-be,* peppering the world with a Babel of biologies. Doyle sees the plurality of possible Artificial Life simulations through a postmodernist lens:

> This paradox at the heart of A-life research—the search for an essence of life through simulation, a simulation with no preexistent referent—is indeed productive. It produces a frenzied spiral of experimentation, performance and debate around the blurred lines between living and nonliving systems. But this is also perhaps a case where a metaphor serves to obscure a more radical question raised by A-life: whether "life" can be characterized in global, "essential" terms, or whether we might better speak of a spectrum or plurality of effects, Artificial Lives. (1992:5)[9]

The Frankensteined-together objects that populate the postmodern world are often a confusing mélange of quotes and references drawn from many different cultural and historical contexts. In Artificial Life, "life"—once an apparently stable and invisible unity that tied together living things of diverse descriptions—becomes a messy mass of qualities that can in various combinations and permutations inhabit organisms, machines, and even entire planets. Following this logic, scientists dedicated to a mechanistic view of the world can begin to entertain Yaeger's "scientific animism," in which minerals and crystals come alive again, as they did in the wondrous worlds of premodern times. In an article entitled "Nonorganic Life," the filmmaker Manuel DeLanda suggests that the sciences of complexity and self-organization of which Artificial Life is a part will allow us to see living process anew. He follows Gilles Deleuze and Félix Guattari who have

> suggested that [the] abstract reservoir of machinelike solutions, common to physical systems as diverse as clouds, flames, rivers, and even the phylogenetic lineages of living creatures, be called the "machinic phylum"—a term that would indicate how nonlinear flows of matter and energy spontaneously generate machinelike assemblages when internal or external pressures reach a critical level, which only a very few abstract mechanisms can account for. (DeLanda 1992:135–136)

DeLanda argues, "In short, there is a single machinic phylum for all the different living and nonliving phylogenetic lineages" (1992:136). The machinic phylum resurrects a great chain of being linking minerals and minds.

Artificial Life holds fast to modernist ideas about the unity of science and the world. But the practice of simulation pulls it along a postmodern trajectory that threatens/promises to undermine these foundations. As Artificial Life seesaws between modernism and postmodernism, generating premodernist spin-offs and echoes, it also undoes the stability of those constructs organized around the signifier "modern." Artificial Life may not be postmodern so much as "metamodern" or "hypermodern," embedded in a process in which "modernity is altered through the amplifications of its own excess rather than being overturned or superceded" (Franklin 1993a:4). But given that Artificial Life is textured by a weave of tales from the history and future of science, religion, politics, economics, gender, sexuality, and race, there is a sense in which Artificial Life could never have been modern, perfectly cordoned off into a realm of eternal asocial rational truth. Definitions of life have never been stable. Following Bruno Latour (1991), we might see Artificial Life as "amodern." As Haraway puts it succinctly, "The amodern refers to a

view of the history of science as culture that insists on the absence of beginnings, enlightenments, and endings: the world has always been in the
middle of things, in unruly and practical conversation, full of action and
structured by a startling array of actants and of networking and unequal
collectives" (1991e:304). On the amodern view, it should not be surprising that Artificial Life is simultaneously an empirical science, an engineering practice, a set of technologies for telling old and new stories
about gender and reproduction, a tool for stabilizing and undermining
existing economic ideologies, and a religious epistemology. Science and
society have never been separated, as modernist mantras would have it.
Science has never described a nature outside human relations, and human relations (or culture) have never completely determined (or constructed) everything that has come to count as nature.

RETHINKING NATURE/CULTURE
AND LIFE ITSELF

Weighing Artificial Life on the scales of modernism and postmodernism
leads to a recognition that the categories we have inherited for making
sense of science are themselves unstable. With this in mind, I turn to an
examination of two categories within the modernist repertoire that are
given a new dizziness by Artificial Life: nature and culture, categories
that—postmodernist vertigo notwithstanding—continue to have a central place in the pivot points of contemporary society.

As I discussed in the introduction, Artificial Life depends in part for
its freshness on a name that sounds oxymoronic at first hearing. "Life"
is supposed to be a paradigmatically "natural" item; the claim that it
might be re-created in another medium threatens to make nonsense of
our idea that a clean line exists between the natural and the artificial.
The oxymoron is a traffic sign, announcing new flows between territories previously understood to be separate. As Marilyn Strathern writes
on Artificial Life, "We are not just supposed to think that machines are
like bodies, but that there are aspects of machines that function no differently from parts of the human body even as human beings may embody technological devices within themselves. The one does not imitate
the other so much as seemingly deploy or use its principles or parts"
(1992b:2–3). Following this logic, "artificial" life can be every bit as
"real" as "natural" life. The "reality" of life does not hinge on its organic origin. Shifting emphasis from the naturalness to the realness of

entities solves for many practitioners the problem of the oxymoron Artificial Life. Artificial life contrasts not with real life but with natural life. Since both artificial and natural life deploy the same principles in their workings, both are real. Larry Yaeger said this to me over email:

> I have mixed feelings about the phrase "artificial life." On the one hand, it is the phrase I first associated with the marvelous new interest that led me to design and develop PolyWorld. It is succinct, and reasonably descriptive. However, due to the vagaries of the English language, I realize that it may forever work against its own best interests by people interpreting "artificial" to mean "not real," rather than "man-made" as Chris Langton originally intended. Since the principal tenet of the field of ALife is that life is a process, not any specific set of materials, it is necessarily the case that "artificial" life can be very much "real."

And Mark Bedau said,

> I think the phrase "artificial life" is fine as a name for the loosely knit interdisciplinary nexus that we are all engaged in. It doesn't seem like an oxymoron, any more than "artificial light" or "artificial grass." As you will note from these two comparisons, I don't think the phrase "artificial life" entails that artificial life is genuine life (like artificial light is genuine light). Whether artificial life could be genuine life is a substantial philosophical question, one that is more or less still open. I'm pretty sure the answer is "Yes," but I have not yet taken the time to work it out.

This shift of balance from "natural" to "real" would seem a departure from the usual epistemologies that have grounded understandings of life in the heaviness of nature. But nature has far from vanished in the epistemology of Artificial Life; in fact, it has been extended to describe all of reality. As Chris Langton has become fond of saying, Artificial Life is itself "natural."

Langton delivered an eloquent speech on this at the fourth Artificial Life conference, held at MIT, where he inaugurated events with a stunning vision of the future of the field. As he sat with a cast-covered broken leg perched on a beat-up chair, he argued that since we humans who make Artificial Life are part of nature, Artificial Life must be natural. Langton displayed a cartoon he had drawn in which he had circled a city skyline. He said, "This is nature," and continued by declaring, "Technology is the current state of nature. We now live with and in 'techno-nature.' " According to Langton, human culture is a form of Artificial Life, and as a part of that nature, is subject to the laws and tendencies of natural systems, at the same time that its dynamics can be used to alter some of those tendencies. He warned of practicing a science of Artificial Life that ignored the "wisdom" of nature. And he cautioned that we must be wary of the Frankenstein syndrome and might rather consider the "Langton-stein" solution, which is to intervene in nature, since we are ourselves part of it, but to do so according to "how ecology would want to behave." In the geographer Neil Smith's terms, Langton wanted two sorts of nature, one as a force external to humans and another as a universal category incorporating "human and nonhuman worlds in endless union" (1996:39). As Langton put it, "Nature is something we want to work with rather than dominate. We should view technology as nature and work toward the naturalization of technology." On this view, nature, while changing as a result of the culture it harbors, ultimately includes and subsumes culture. Strathern diagnoses such a view well in her description of the Euro-American sense of nature and culture: "While at one level, a contrast between the natural and the artificial might distinguish different views of culture, it might equally distinguish Culture itself, as intrinsically artificial, from Nature, the source of all that [is] natural. Cultures, in this European view, [are] artificial creations natural to the human condition" (1992b:48).

Against this notion that the cultural project known as Artificial Life is natural, we could argue that the "nature" that Artificial Life researchers seek to imitate and locate their project within is saturated by the categories of "culture," or more narrowly, the categories of a *particular* culture or set of cultures. We might see Artificial Life in close company with genetics, immunology, and environmentalism—those discourses Paul Rabinow has designated "the leading vehicles for the infiltration of technoscience, capitalism and culture into what the moderns called 'nature' " (1992:245). While this position risks reifying both culture and nature, it is only a slight caricature of the position I myself held during a talk I gave at the MIT Artificial Life conference. An account of this talk supplies a sense of the dance in which the concepts of nature and culture are locked, as well as a sense of the choreography of my engagement with the Artificial Life community.

My talk traced in tourist-paced form the territory I explore in chapters 2 and 3 of this book. I gave my talk directly after Tom Ray had given his solicitation for an Internetwide version of Tierra. When I finished reading my prepared text, I was interrogated by the audience, who asked me a good hour's worth of questions (my talk was at the end of an afternoon of speaking events). At moments when I was foundering in my attempts to answer more challenging or confrontational questions—especially questions about why my story was so different from practitioners' accounts—Langton graciously intervened, defending my right to discuss Artificial Life scientists' work using a language not always their own: "An anthropologist, if he was studying a group of people in the Amazon, would not necessarily report his findings in the context of their cosmology. I think it's perfectly legitimate for him to have his own distinct, descriptive venue." Langton also drew attention to the ways interdisciplinary ventures often require arguing about language. When my inquisitions were through, I wandered around, waiting for the evening's Artificial Life clam bake to begin. As people began to accumulate at tables on the grass, a few approached me. Some were amused, some supportive, some confused, and some hostile, taking time to chew me out for misunderstanding the scientific method and for bringing a feminist analysis to a subject they emphatically felt had no gendered dimensions.[10] Tom Ray in particular expressed dissatisfaction with my representation of his work and when he read my dissertation, asked that he have a chance to respond to my arguments when the book was produced. As this book went to press, he composed the following comment, based on his reading of one of the final drafts of chapter 3:

> I find Helmreich's work to be a poor example of anthropology, because rather than trying to understand his subjects, he forces them into the template of anti-heroes in his own politically motivated world view. We could be replaced by anyone and his thesis would be essentially the same. Helmreich advances his own political agenda, the character of which is revealed by the literature that he copiously cites. His analysis has the feeling of a societal equivalent of a Freudian analysis. Rather than our behaviors being determined by hidden events in our infancy, he asserts that they are determined primarily by obscure events in the history of our culture, long before we were born. He devotes several pages to what he sees as the politically incorrect meanings and uses of the word "seed," meanings which he claims are inextricably and historically tied to the work and workers of Artificial Life. But Stefan never learned the meaning of "seed" which is in my mind. For his purposes he didn't need to know. Rather, he introduces a medieval image of rotten semen incubating in horse dung as though it somehow reveals something about his subjects rather than himself. I am tempted to conclude that Helmreich has not learned from his predecessor, Gusterson, who concluded that how researchers use language is "ultimately unconvincing as a key to their inner psychology" (to quote Helmreich's own summary of this position). As a result, Helmreich's book is more about the world according to Stefan, his own mind and world view and the politically incorrect demons that haunt him, than it is about his putative anthropological subjects. It is sad that anthropology has come to this.

Ray's response speaks to the discomfort several researchers expressed to me about cultural accounts of science. Because such accounts spotlight the broad cultural contexts in which sciences exist, they often gloss over the diversity of individual motivations and meanings; moreover, because a variety of stories can be told about what makes a science like Artificial Life function, such stories often come across as mere opinion. My story certainly expresses my opinions, but these opinions are formed in alliance and conversation with others, including Artificial Life scientists. Because I saw myself personally and culturally implicated in Artificial Life, I not only sounded people out on their views but also challenged their ideas, and was myself persuaded of biological, philosophical, and political positions of which I had initially been skeptical. My analytical impulses were governed by a desire to see Artificial Life from outside the most powerful frames of reference, to examine the stakes in describing things in some ways and not others, and to take seriously the idea that language and culture can inhabit us in ways that exceed our intentions. That there was space for an anthropologist at the conference at which I spoke indicated the community's recognition of itself as a cultural entity, as a field with a history that could have been otherwise. Artificial Life and anthropology are both enmeshed in modernity's vexed project of self-examination and self-critique, and both are in transformation as their objects of investigation—"life" and "others"—are under stress as coherent categories. Langton's early study of anthropology is a pointer to shared histories.

Latour (1991) has argued that maintaining that all knowledge is socially constructed is just the mirror image of appealing to nature as the ground for all knowledge claims. From a social constructionist point of view, I could argue that Artificial Life and its images of nature are determined by culture. Artificial Life researchers could nod in agreement and then say that this culture is itself part of nature, that culture, understood as a system of information, is itself a form of artificial life. We could continue in this cycle, with some researchers veering into arguments about how Artificial Life reveals that our understandings of life are constructions. For all of Latour's arguments that nature and society occupy symmetrical positions in this language game, however, I think arguments from nature have been generally more influential in the world of Judeo-Christian Western science. The wheel will almost always cease its cycling when nature is ascendant. Nature remains a powerful reference point for arguments about what is moral or inevitable precisely because one can easily lodge culture in its embrace. As the anthropologist Anna Tsing notes, "Although cultural analysts over and over demonstrate the cultural shaping of so-called natural attributes, we can never thus unseat 'nature': It is an aspect of nature to be partially and ultimately wrongly labeled by human cultural efforts; this is what gives nature its majesty" (1995:114–115). For this reason, I have been prone to err on the social constructionist side rather than the natural empirical side.

In "Romancing the Helix" (1995) the anthropologist Sarah Franklin contends that as cultural and natural logics bleed into one another—especially as each are distilled into information-processing systems—attention has shifted from understanding the facts of nature to understanding the secrets of "life," where life is an object or process that incorporates both natural and cultural elements. Various technologies of assisted conception now allow people singly, in couples, and in groups to engineer the fusion of gametes, to manipulate zygotes and blastocysts in the lab, and to contract out or into the gestation of embryos and fetuses. New reproductive technologies—including in vitro fertilization, cloning, surrogacy, embryonic gene therapy, and postmortem maternal ventilation—render problematic the natural status of conception, gestation, and birth and fix attention on "life" (see Strathern 1992b; Franklin 1993a, 1993b, 1997; Casper 1995; Clarke 1995; Hartouni 1997). In the age of the Human Genome Project, practices like the revival of 30-million-year-old dormant bacteria, the manufacture of transgenic plants and animals, and the genetic modification of organisms in line with various cultural preferences also muddy the waters between the technological and the organic. Recent discussion of "biodiversity" as a resource to be culturally recognized, harvested, and managed has focused attention on a cultural and natural "environment" rather than "nature" (see Escobar 1995). As plant and animal genes that produce pharmaceutically interesting compounds are bar-coded, taxonomized, and, often, patented, genes come to condense both natural and cultural logics (see Rabinow 1992; Strathern 1992b; Haraway 1997; Hayden 1998). And if genes are often considered synonymous with life, then, at the end of the twentieth century, we are witnessing a shift from a fetish of asocial nature to a concern with socialnatural life, or even, as Franklin (1993a) has termed it, with "life itself," life considered as an abstract process.[11] This concern with life can be good or bad. Good, if it alerts us to our responsibilities in fashioning life as a process we inhabit personally and politically. Bad, if it freezes life as a thing in itself, occluding the dynamics of objectification and commodification that assist in materializing it in this way. As Haraway writes, "Life itself depends on the erasure of the apparatuses of production and articulatory relationships that make up all objects of attention, including genes, as well as on denial of fears and desires in technoscience" (1997:144).

This new concern with "life" was apparent in the comments of some researchers I interviewed, who thought the debate about whether Artificial Life was real was missing the point. Ken Karakotsios emailed me this reflection:

> If one said artificial life was life of non-biological origin, and that's how it's different than us, many people would probably agree. But that's exactly wrong. Artificial life means literally man-made life. Man-made life is really nature-made life, as man is part of nature. It's still life creating life. "In the beginning," the life processes had to start up one by one from a situation of nonlife. So when we trace our chain all the way back, we ourselves have non-biological origins. That somehow sounds "artificial." So the chain of reasoning conflicts with our basic intuition. . . . I think it's good to consider the oxymoron, because what it really tells us is that life is continuing its business, but it's undergoing a sort of phase transition. We've been in a state up until now where all life has been built upon one thread—wetware. Now we're spreading life out into multiple threads. We're creating diversity in a whole new dimension. Life is extending the dimensionality of its own space. . . . Life not only evolves, but it changes the ways in which it can evolve.

I use the figure "life" to organize some final thoughts on possible futures that Artificial Life may index, diagnose, enable, or forestall. There are many possibilities, but I focus on two extremes, one in which normative ideas about reproduction are reproduced in new and more powerful ways and another, perhaps queerer, possibility in which Artificial Life practices of simulation undermine dominant senses of what is natural, cultural, and essential. I have played with Langton's terms *life-as-we-know-it* and *life-as-it-could-be* in titling these last two sections.

LIFE-AS-WE-KNOW-IT:
REPLICATING REPRODUCTION IN ARTIFICIAL LIFE

In "The Work of Art in the Age of Mechanical Reproduction," the cultural theorist Walter Benjamin wrote, "Every day the urge grows stronger to get hold of an object at very close range by way of its likeness, its reproduction" (1936:223). That object that Artificial Life researchers want to get hold of is life, understood for computationalists of the SFI breed as reproducible in its essence, since its essence is considered to be a replicating information structure. For Artificial Life scientists who hold that life resides in reproduction, replicating reproduction in computers will grant them the grape for the Grail of Artificial Life: a self-replicating program that mimics vitality so well that it transubstantiates into vitality itself.

Benjamin argued that when a work of art is mass reproduced, it enters into a sphere of existence that robs it of its uniqueness, its authenticity, its "aura." Following Benjamin's romantic impulse, we might say that when "life" is dittoed *in silico,* it enters into a sphere that robs it of its authenticity and essence. Parting company with the nostalgia of Benjamin and the empiricism of Artificial Life scientists, though, we might also say that it enters into contexts that reveal that there never was an "essence" of life.[12] As Doyle notes, "What makes possible the substitution of the signs of life for life is the reproducibility of 'lifelike behavior,' a reproducibility that ultimately points to the fact that A-Life organisms are themselves reproductions, simulations cut off from any 'essence' of life" (1997b:122). Digital organisms become forgeries of already chimerical originals. Continuing this curve of thinking, we could say that when reproduction is reproduced in computers, we learn that what we call reproduction emerges out of cultural and scientific definitions and practices. We learn that calling it reproduction itself indexes commitments that make thinkable its duplication in computers.

Many Artificial Life scientists see themselves ushering in a new stage of evolution, one in which novel life-forms will be birthed through scientific conceptions leading to self-reproducing computer programs. They view themselves as kinds of double agents, with biological reproduction and the machinic reproduction they are engineering as geminate strategies in a larger evolutionary game. They see themselves as "in the employ" of evolution, creating new life-forms that will unchain themselves from carbon chemistry, perhaps traveling off-planet in robot-bodied silicon splendor. Artificial Life researchers can claim that organic biological reproduction can be subsumed, transcended, and devoured by new technobiological reproduction because they use the word *reproduction* to refer to both the perpetuation of practices and ideas and the generation of new organic beings (see Harris and Young 1981). This is how Artificial Life has become thinkable.

Strathern has commented that the Euro-American reproductive model "makes us greedy for both change and continuity, as though one could bring about momentous (episodic) change while still being regarded as the continuous (evolutionary) originators of it" (1992b:177). These words illuminate the cultural logic beneath Artificial Life researchers' contentions that the manufacture of artificial life is both novel and evolutionarily inevitable. When Langton writes, "The creation of life is not an act to be undertaken lightly. We must do what we can to ensure that the future is equally bright for both our technological and our biological offspring" (1992:22), we learn that reproduction is the real fuel for Euro-American time travel into the future. Langton (1989:2) has suggested that Artificial Life will expand biology's purview to include not just *life-as-we-know-it* but also *life-as-it-could-be*. Given this extraordinary charter, it is surprising that Artificial Life stories about generation are so commonplace, that they reiterate all-too-familiar stories about monogenetic masculine creations of life, that they are stories about "reproduction," rather than, say, rhizomatous regeneration. Artificial Life scientists' imitations of nature often tend to be recreational vehicles into dominant nostalgias for the natural, into places populated with *life-as-it-should-be*.

There are of course those in Artificial Life who see the discipline fixing attention on the artifactual character of our visions of nature. We might see Artificial Life alongside new reproductive technologies as a practice in which fragments and foundations of the natural are both stabilized and troubled, in which ideas about the given and inevitable become newly contestable. Artificial Life resurrects old ideas in new computer codes, but as Strathern reminds us, novel twists on received texts are always possible: "The ideas that reproduce themselves in our communications *never reproduce themselves exactly*. They are always found in environments or contexts that have their own properties or characteristics" (1992b:6). It is to a possible twist in the reproductive logics of Artificial Life that I dedicate the following twin section.

LIFE-AS-IT-COULD-BE: ARTIFICIAL LIFE AS DRAG PERFORMANCE

The mimetic promenades of Artificial Life programs can be read in two ways: either they affirm that life is a sturdy, natural category amenable to better or worse imitation, or they undermine this idea. Artificial Life depends on endorsing both positions simultaneously. The first provides practitioners with an object to copy, and the second offers hope that their replicas can transform into new life. This makes the question of how objects are alive not a question of whether they are natural but of whether they perform as though they were alive. Doyle frames the tension as follows:

> Given that the genetic problem of A-life is the lack of an adequate definition of life, whence comes our criteria for selecting the proper objects of artificial life from the pretenders? The answer is, of course, in practice. The performance of A-life, given the lack of immutable criteria for "life," relies on its rhetorical software for its plausibility and its fascination. "An automata is not just a machine, it is also the language that makes it possible to explicate it." This language, metacode, or rhetorical software, despite Langton's goal of finding the "essence" of life, in fact works toward another goal, the evaporation of the difference between living and nonliving systems. . . . In the absence of any adequate pregiven definitions of life, the plausibility of A-life creatures rests on their ability to simulate something for which we have no original, "lifelike behavior." (1997b:121–122)

The vanishing of life and nature over the horizon of endless simulation heralds a disturbance in how we fashion the real, and I want to press on some of its implications by proposing that we understand Artificial Life simulation through another variety of performativity, namely drag performance, where drag performances—cross-dressing from one gender to another—are impersonations that both dismantle and consolidate the natural status of the object under imitation. Just as male-to-female drag mimics normative images of femininity, either reifying femininity or making us aware that our gendered identities are performances (albeit often painful and involuntary ones), so does Artificial Life simultaneously solidify and liquefy our conceptions of life.[13] In Artificial Life, computer programs are made to cross-dress as organisms. Recall Helena's question: "Would you say that words on a page are alive just because they're information? I wouldn't. They might be able to convey the sensation of life to someone who reads them, and that's what this [ALife] is doing. This is giving an impression of life. It's like, Here's a computer's impression of life." The computer's impressions of life are reproductions that assure some people of the old-fashioned facts of life, but we might well see these likenesses in a different light. We could ask, with the cultural critic Marjorie Garber, who has written extensively on transvestism and transsexuality, "What, then is the relationship between transvestism and repetition? For one thing, both put in question the idea of an 'original,' a stable starting point, a ground. For transvestism, like the copy or simulacrum, disrupts 'identity' and exposes it as figure" (1992:369). All we need to do to make this work as a comparison with Artificial Life is to rewrite as follows: What, then is the relationship between Artificial Life and simulation? For one thing, both put in question the idea of an 'original,' a stable starting point, a ground. For Artificial Life, like the copy or simulacrum, disrupts 'nature' and exposes it as figure.[14] Calling Artificial Life a kind of cross-dressing is meant as a deliberate insubordination, one that I hope can drag this scientific practice into the realm of the unruly cultural, into the realm of play, but play with serious consequences for how we think about our lives. Let me gesture toward these consequences with a story about a game.

There is a founding moment in the history of computer science that is repeated again and again in introductions to Artificial Intelligence. In the 1940s, Alan Turing proposed that computers might one day be said to think, and he proposed a test for determining how one might tell when computers had attained sentience (see Turing 1950). The "Turing test" consisted of an imitation game in which an interrogator communicated simultaneously with a computer and a person over teletype. The interrogator was allowed to ask any range of questions he or she liked, all in an effort to see if the computer could respond in such a way as to be indistinguishable from a person. If the computer passed as a person, it could be considered intelligent.

Most narrations of the Turing test forget that Turing introduced his game as a twist on an old parlor game having to do with gender and mimesis. Turing's first phrasing of the test had a man (A) and a woman (B) trying to persuade an interrogator that they were both women. Turing turned this into a computer science game when he asked, "What will happen when the machine takes the part of A in this game? Will the interrogator decide wrongly as often when the game is played like this as he does when the game is played between a man and a woman?" (Turing 1950:2100). As Bill Maurer points out, the Turing test effectively asks: " 'Can a machine gendered male successfully enact being a woman?' or 'Can a "male" machine be a convincing transvestite?' " (1992:7). At one of the primal scenes for Artificial Intelligence—scripted, we should note, by a gay man who struggled tragically with his own relation to gender and sexuality (Turing committed suicide in 1954 while in a U.K. prison serving time for being homosexual)—we have computer performance wrapped up in a project of imitating categories of behavior that many see as affixed to the "natural" category of sex/gender. We might conclude that if a computer can imitate these characteristics, there is nothing essential or natural about them to begin with.

The feminist theorist Judith Halberstam has made a similar argument and maintained that Turing's use of the sexual guessing game demonstrates an

> obvious connection between gender and computer intelligence: both are in fact imitative systems, and the boundaries between female and male . . . are as unclear and as unstable as the boundary between human and machine intelligence. . . . Gender, we might argue, like computer intelligence, is a learned, imitative behavior that can be processed so well that it comes to look natural. Indeed, the work of culture in the former and of science in the latter is perhaps to transform the artificial into a function so smooth that it seems organic. (1991:443)

I suggest that life is also a set of behaviors amenable to imitation, and as such does not consist of a set of natural properties. Entities come to life as they are described and built in ways that reiterate, reference, or cite norms of behavior classed as belonging to the living, just as we become gendered as we are compelled to fit into masculinities and femininities we did not ourselves create. Neither life nor gender are natural categories. Figure 22, art for an early Artificial Life video called *Stanley and Stella, in "Breaking the Ice,"* shows how computer graphics that attempt to reproduce the signs of life and gender necessarily call attention to the techniques that make such categories seem natural. Stanley the boxy manbird and Stella the curvaceous ladyfish are drag performers as well as icons of Artificial Life.

This is not to say that "life" and "gender" do not have real, pleasurable, and painful effects on our experiences of ourselves and others. Gender is often treated as the natural consequence of "biological" sex identity, which becomes an anchor point for all sorts of medical specifications and enforcements of who we are. One's "sex" is an effect that is produced in one's body, but it is not ultimately natural because of this. We might say for life what Butler has said for sex:

> The category of "sex" is, from the start, normative; it is what Foucault has called a "regulatory ideal." In this sense, then, "sex" not only functions as a norm, but is part of a regulatory practice that produces the bodies it governs, that is, whose regulatory force is made clear as a kind of productive

Figure 22.   Still from *Stanley and Stella, in "Breaking the Ice."* Courtesy
Symbolics, Inc., and Whitney/Demos Productions.

power, the power to produce—demarcate, circulate, differentiate—the
bodies it controls. (1993:1)

There is no ultimate ontology of life—which does not prevent it from
materializing as a process that shapes entities. And Artificial Life is dedicated to printing life's second edition in machines. But an attention to
drag allows us to see this in a new way. Viewing Artificial Life as drag
affords a sense that what counts as life is a contingent social accomplishment. Artificial Life need not be about telling the same old neo-
Darwinian stories. Other tales can be told using the technologies of simulation. Artificial Life could be seized as a truly useful tool for imagining
life-as-it-could-be. While my starting point is different, I agree with most
researchers in the field: Artificial Life *can* be a provocative new way of
thinking about the life sciences.

But viewing Artificial Life as drag performance alerts us to how mimesis always happens in the frame of existing power. My own choice of
male-to-female drag as an example highlights how this is so. In many

contexts, male-to-female drag can stabilize heterosexual structures of desire in a deeply misogynist way (Butler would point us to movies like *Victor Victoria* and *Tootsie*). As Butler tells it, drag

> serves a subversive function to the extent that it reflects the mundane imper-sonations by which heterosexually ideal genders are performed and natural-ized and undermines their power by virtue of effecting that exposure. But there is no guarantee that exposing the naturalized status of heterosexuality will lead to its subversion. Heterosexuality can augment its hegemony *through* denaturalization, as when we see denaturalizing parodies that re-idealize heterosexual norms *without* calling them into question. (1993:231)

At many moments in drag, then, nothing is undermined; rather, the ideal copied is reified as stable—and this quite apart from whether it is believed to be natural (we might think also of the practice of "black-face" in white America, in which African-Americans were caricatured in ways that only stabilized what many whites presumed were multiply unbridgeable differences between themselves and blacks). Artificial Life is produced by social subjects in positions of relative power, and, for this reason, the mimeses it produces might well become potent tools for reinforcing dominant ideas about life, along with the gendered, raced, classed, and sexualized imaginaries these encode. In spite of this, I still choose drag as a vehicle for envisioning more transgressive possibilities in Artificial Life because, as Butler puts it, "performativity describes this relationship of being implicated in that which one opposes, this turning of power against itself to produce alternative modalities of power, to establish a kind of political contestation that is not a 'pure' opposition, a 'transcendence' of contemporary relations of power, but a difficult labor of forging a future from resources inevitably impure" (1993:241).

In 1950, Alan Turing posed the question, "Can a machine think?" and immediately observed that the answer depended on how the words would come to be used in practice. He wrote, "I believe that at the end of the century the use of words and general educated opinion will have altered so much that one will be able to speak of machines thinking without expecting to be contradicted" (1950:2107). As the year 2000 comes and goes, there are few who have changed their definition of thinking such that it now includes computers.[15] Even IBM's chess-play-ing Deep Blue has been disqualified from cognition, understood as a complex calculator rather than a chip off the old block of human mind.

But there do seem to be those who are changing their definitions of "life" such that they are comfortable claiming that computers could come alive or could support living digital creatures. In the silicon second natures of Artificial Life, we can discern possible futures for how life, nature, and culture may be reproduced, recombined, and, perhaps, productively reconfigured.

Coda

AS THE WRITING of this book wound to its tail, I purchased my first virtual pet, a Tamagotchi, a matchbox-sized piece of artificial life from Japan, possessed of a digital body visible as a collection of dots on a small screen set in an egg-shaped plastic pendant. Sitting in a cybercafe in New York City some three years after my primary fieldwork, listening to my Tamagotchi bleep for attention, I thought about the infiltration of Artificial Life techniques and metaphors into the everyday workings of machines. The words of one Institute scientist came back to me: "As computers become more sophisticated, they have more and more properties in common with living things. I really do believe that in the next five years people are going to start looking at computer systems as much more biologically oriented. You see this already with computer viruses—and that's just the first step."

Some of the next steps have produced such entities as "knowbots," artificially evolved software agents that can be trained to search the Internet for information specific to the desires of individual users. The Internet itself has been populated by a bevy of digital life-forms, with a rich mesh of Artificial Life worlds increasingly online. The Contact Consortium, a northern California nonprofit organization incorporated in 1995, has devoted itself to building Internet-accessible three-dimensional digital realities in which humans and artificial life-forms can interact. One of their special interest groups, Biota.org, seeks to provide through the World Wide Web a suite of virtual environments (delivered via VRML, the Virtual Reality Modeling Language) in which "geographically separated students, teachers, researchers and the interested public can meet . . . to assess the results of experiments or design new ones, . . . using a flexible and open interface to enable students to modify and share genetic structures, rewire behavioral models and set the parameters of the ecosystems they bring into being" (*www.biota.org/grants/index.html* as of January 28, 1997). Because user participation is encouraged in Synthetic Ecosystems, there is a promise that the closed worlds of Artificial Life may become more open and unruly as programming goes polytheistic. Biota.org's educational mission is animated by an extraterrestrial dreamwork common in Artificial Life; its webpage—a kind of parallel cousin to the spaced-out site once maintained by the ascended faithful of Heaven's Gate—tells us that life is pressing to be elaborated into digital forms that may eventually leave their cyberspace cradle and transmit themselves at light speed through outer space, finding "a way through the keyhole of human technology, escaping both the bounds of Earth and the mortal coils of the double helix" (*www.biota.org/org/vision.html* as of November 30, 1996). The Synthetic Ecosystems project draws inspiration from Tierra, PolyWorld, and the Sugarscape, and participation from computer scientists, biologists, virtual community experts, artists, and businesspeople. In August and September 1997, Biota.org sponsored a gathering entitled "The Burgess Shale and the Digital Cambrian: A Conference on the Origins and Future of Life on Earth," held in Banff, Alberta, Canada, near the Burgess Shale, one of the best-preserved fossil records of the Cambrian period. The reference to the digital Cambrian drew directly from Tom Ray's proposal for a global Tierra. Indeed, Ray has been an important ally of Biota.org.

Ray has gone forward on Network Tierra, but differently than he originally proposed. Finding people reluctant to make space for Tierra in their Internet accounts, he has gathered nearly 150 accounts of his own, at various locations around the world, setting up a collection of nodes he sees as akin to a string of islands. Ray has made Tierran organisms more "net savvy," giving them the ability to "surf" to other nodes on the Tierra network (a space within the Internet that Ray compares to the Web). The reaper has gone global, too: an "apocalypse function" periodically kills creatures that stay still too long.

Ray himself has moved to another node in the worldwide web of Artificial Life. Since autumn 1994, he has been working in Kyoto, Japan, at the ATR Human Information Processing Research Laboratory, a company dedicated to research in telecommunications and human/machine interaction. Scientists in Japan have been watching the development of

Artificial Life with interest and have recently organized their own conferences and research groups. In May 1996, the fifth conference on Artificial Life was held in Nara, Japan, marking the movement into an Asian arena of a series of conferences that had so far unfolded in the United States (in Santa Fe and at MIT). For many of the U.S. and Europe-based scientists I spoke with, Japanese interest in Artificial Life signaled the motion of the field into fast-forward. In mainstream European and American imaginations, Japan is a land of the future, a place where technological innovation happens quickly, aided by well-coordinated national initiatives in computer science and microelectronics. When I traveled to Artificial Life V, I saw Western researchers working to understand how familiar ideas were being developed in a context organized by different research imperatives and priorities.

The conference opened in an auditorium containing a stage fashioned after a Noh theater space, with a wooden roof and a surrounding stone garden moat. The plenary talks introduced one of the novel themes to emerge at this gathering: the fusion of microelectronic technology with organic materials. In a talk entitled "Insect-Model Based Microrobot," Hirofumi Miura reported on hybrid automata built using robot and insect parts. One device consisted of a wheeled microrobot capable of following pheromone trails using sensors built out of the antennae of a real male silk moth. Another joined real cockroach legs to a paper body: "Legs can be actuated by appropriate electrical stimulation and the robot can walk" (Miura et al. 1997:29). Located in the Department of Mechano-Informatics at the University of Tokyo, Miura's group also constructs 1-millimeter insectlike robots with flapping wings made of silicon wafers folded in an origami style (Miura et al. 1997:28). Extending the discussion of cyborg systems, the computer scientist Hiroaki Kitano argued in "Towards Evolvable Electro-Biochemical Systems," "There are chances that biochemical process can be incorporated in electric circuits as a part of very large integrated circuits.... The most salient characteristics of such a computing system is that it can incorporate metabolism of both information and matter.... By using biochemical devices, the distance between natural and artificial biological systems can be reduced substantially" (1997:219). This mixture of the organic and the synthetic gives new meaning to silicon second nature, fleshing out a different kind of communion with the silicon wafer than that afforded by computer simulation. Electrobiochemical artificial life-forms are neither virtual nor transgenic. Mixing signals from nature and culture, they are transinformatic, the result of a new sort of AI: Alchemical Informatics, not Artificial Intelligence.

Though these materially new creations captured the most attention at the conference, the floating worlds of simulation retained their popularity. New spaces for virtual life were in abundance, as were a variety of interactive artworks made with techniques developed in evolutionary computation. In one popular presentation, the artist Naoko Tosa and the computer scientist Ryohei Nakatsu of ATR discussed computer-generated facial images that could be trained to respond to user input. Linked to neural nets, these faces developed a repertoire of expressions in response to users' vocal modulations. One net was named MIC, a baby boy character able to laugh, cry, and throw tantrums: " 'MIC' is a male child character. He has a cuteness that makes humans feel they want to speak to him. He is playful and cheeky, but doesn't have a spiteful nature. He is the quintessential comic character" (Tosa and Nakatsu 1997:143). As a new artificial life entity, MIC is a kind of hyperactive, hypertext Tamagotchi. Tosa argued that one application of MIC might be in cross-cultural cyberspace communication between Western and Japanese researchers; people might augment real-time written communication with MIC images conveying the emotional tone of their words. Users would join MICs as trainees in the apprehension of different people's moods.

Because Japan and the West are so persistently posed as cultural opposites—even opponents—in popular accounts of technoscientific innovation, cross-cultural communication has been a watchword for people wishing to build transpacific social and economic ties. Announcements for Artificial Life V urged attention to bridging cultural difference as Artificial Life traveled from "the land of enchantment" (New Mexico) to the "land of the rising sun" (see Langton 1994). At the gathering itself, the philosopher Brian Keeley and the historian of science Osamu Sakura led a panel entitled "What Is Life? West and East," a moment of explicit reflection on the influence of cultural tradition in science. But while the panel promised an evenhanded if somewhat caricatured discussion of difference, most conversation centered on the possibility that Asian mystical traditions might shape conceptions of "life" in Japan, paying little attention to the metaphysical commitments packed into "life" in the West. For both Japanese and Western discussants, Japan emerged as the entity with "culture." One Japanese panelist expressed skepticism that computers or robots could be alive, observing that they lacked what the

"East" called "Chi," that vital energy that interacts with external elements, that gives life its movement.

At the same time, the conference saw many presenters detail a Japanese history for the practice of synthesizing vitality. The historian Shouji Tatsukawa demonstrated a replica of an eighteenth-century Japanese tea-serving robot. Some Japanese scientists said that Artificial Life had resonances with Japanese traditions of rearranging nature—in gardens and in the art of flower arrangement. Others said that microelectronics grew from a cultural tradition of miniaturizing nature, as exemplified by the practice of bonsai (indeed, a virtual reality bonsai tree was featured at one presentation). Such practices were taken to typify a tradition of combining the natural and artificial, a tradition carried forward in such objects as mixed-media microrobots.

Narratives of the inevitability of Artificial Life in Japan are, of course, just as invented as those produced in the West. Both submerge histories of transnational connection in art and technology, and both ignore the recency of scientific concepts of "life." It should be said, however, that Japanese scientists who presented such clean narratives were advancing a complicated agenda. They were reacting in part to Western stereotypes of Japan as a nation of competent and competitive copyists, as a nation that, as one Westerner I spoke with put it, "never originated any really new technology." Where many Americans and Europeans saw Japan appropriating Western technoscience and speeding it into the future, Japanese speakers linked their interest in Artificial Life to established traditions. In an elegant observation connecting the traditional and the novel, the cultural anthropologist Noriyuki Ueda remarked that he could think of no better setting for a discussion of Artificial Life, West and East, than the Noh theater space in which conversations had taken place. Noh theater circulates around ghosts, beings on the border between life and nonlife.

One notable absence at this conference was Chris Langton, who was unable to attend but who sent messengers of a new project he was working on in Santa Fe. Recently, Langton and a team of SFI researchers have been fashioning a general-purpose simulation tool called Swarm, "a multi-agent simulation software platform for the study of complex adaptive systems. In the Swarm system the basic unit of simulation is the *swarm*, a collection of agents executing a schedule of events. Swarm accommodates multi-level modeling approaches in which agents can be composed of swarms of other agents in nested structures."

In a typically inventive linguistic turn, Langton has dubbed Swarm "SimSim." In a typically clever commentary, Richard Doyle remarked to me that this neologism promises a science not just of *simulation-as-we-know-it* but of *simulation-as-it-could-be*. This nesting of possibility within possibility is the name of the game in the age of simulated second natures, a sign that we may be achieving escape velocity from science as usual, traveling toward recognizing the multiplicity of scientific ways of knowing. At Artificial Life V, David Ackley spoke for the plugged-in digerati when he announced, "Software is already alive. The question is not how to make artificial life, but what kind of artificial life do we want?" Dancing frenetically on its small screen, my Tamagotchi tells me that Ackley's declaration is both premature and prescient. Ackley's question is more pressing, for it wires us into crucial issues of possibility and responsibility in a world increasingly circuited by the objects and logics of synthetic vitality, of artificial life.

Notes

INTRODUCTION

1. Describing the Institute as "private" doesn't tell the whole story. Many of its projects are funded by federal agencies like the National Science Foundation and the Department of Energy.

2. Other significant Artificial Life research groups exist in the United States at MIT, the University of Michigan, UCLA, the California Institute of Technology, and Stanford University and in Europe at the École Polytechnique in Paris, France; the Université Libre de Bruxelles/Vrije Universiteit Brussel in Belgium; the Universidad Complutense de Madrid in Spain; the Universidad del País Vasco/Euskai Herriko Unibertsltate in San Sebastián, Spain/Donostia, the Basque country; the University of Sussex in Brighton, U.K.; Bioinformatica in Utrecht, the Netherlands; the Institut für Theoretische Chemie der Universität Wien in Austria; and the Istituto di Psicologia, Consiglio Nazionale delle Ricerche in Rome, Italy. There is a growing interest in Artificial Life in Japan, especially at ATR Human Information Processing Research Laboratories in Kyoto and at various electronics companies.

3. There is another species of Artificial Life, somewhat less publicized, called "wet" Artificial Life. This involves the directed molecular evolution of such biomolecules as RNA, a practice of great interest to pharmaceutical companies looking for new sorts of designer drugs.

4. For a less Santa Fe-centered history, see *Vida artificial,* by the computer scientists Julio Fernández Ostolaza and Álvaro Moreno Bergareche (1992), or its updated French translation, *La vie artificielle* (1997).

5. During my research, I interviewed forty-four Artificial Life researchers (mostly people who identified themselves as such) and ten scientists working in other branches of the sciences of complexity, such as evolutionary economics or the physics of information. I grilled a few of these people two, three, or even four times. I conducted most interviews at the Santa Fe Institute, though I did several at conferences or over electronic mail. Formal interviews were supplemented by

casual conversations over meals, at parties, in corridors and lounges, and during official social functions. I interviewed a diverse set of researchers to discern differences and similarities between people by disciplinary affiliation or origin, university of graduate training, country of scientific training and/or employment, theoretical orientation, ethnicity, gender, sexual orientation, generation, political affiliation or commitment, and religious belief or upbringing. My interviews began with questions about researchers' specific work and about how, why, and through what connections they became involved with Artificial Life and/or the Santa Fe Institute. I asked about programming, institutional politics, the relationship of personal life to scientific work, philosophical and literary inspirations, national and international contexts of Artificial Life research, possible applications of Artificial Life technology, and the future of the field. I asked how people regarded the work of colleagues and what they considered significant trends and controversies. I also interviewed staff at the Santa Fe Institute and some of the partners and friends of researchers.

I supplemented information from interviews with participant-observation. I spent time with people as they thought through and implemented Artificial Life systems, and I spent uncounted hours wrestling through simulations myself, calling on my anthropological background in population genetics and evolutionary biology and my familiarity with the programming languages of Basic, Pascal, C, and LISP. Conversing in the language of biology and computer science was essential in following the microprocesses of programming. Interacting with researchers informally helped me to understand what acceptable scientific talk sounded like, under what conditions people were allowed to bring in personal anecdotes to motivate their speculations, and how scientific ideas were pitched, stabilized, dismissed, or turned into research programs.

At international conferences, I got a flavor of how the Santa Fe Institute was perceived in the world arena. Subsequent to the first conference I attended in Paris, I attended Artificial Life III in Santa Fe in June 1992; the second European Conference on Artificial Life in Brussels, Belgium, in May 1993; a small European meeting on Artificial Intelligence and Artificial Life in San Sebastián, Spain/Donostia, the Basque country, in December 1993; Artificial Life IV at MIT in July 1994; and the third Conference on the Simulation of Adaptive Behavior in Brighton, U.K., in August 1994. I also attended the third European Conference on Artificial Life in Granada, Spain, in June 1995 and Artificial Life V in Nara, Japan, in May 1996.

Finally, I read extensively in the Artificial Life literature, which includes conference proceedings, articles in journals, Institute article preprints, informally distributed papers, and a store of professional digests available over the Internet. Reading these literatures helped me fine-tune interview categories and immerse myself in current debates.

6. There are a few other cultural theorists who have taken a direct interest in Artificial Life. This book is in dialogue with Lars Risan, an anthropologist from the University of Oslo in Norway who worked with evolutionary roboticists stationed at the University of Sussex. Risan (1996) attends to how rhetorics of complexity and computational flexibility in Artificial Life might allow for an

ecological and situated sense of the observer's embeddedness in systems. This book is also in conversation with Richard Doyle, a rhetorician examining how notions of "life" have mutated under the sign of information (see Doyle 1997b). Literary theorist N. Katherine Hayles's work on virtual and artificial worlds, and on Tierra in particular (1994a, 1994b, 1995, 1996), has also been crucial for my thinking, as has historian Julian Bleecker's (1995) work on the semiotics of simulation games.

7. For the moment, I am bracketing views of nature as unruly, wild, and irrational, as a force that morality must tame and protect against.

8. Yanagisako and Delaney maintain that the universal nature described by Western science inherits many of its traits directly from monotheistic cosmology: "In the nineteenth century the Biblical worldview began to collapse and God began to drop out of the picture—at least among certain members of the intellectual elite—and what was left was a rule-governed Nature, Nature stripped of its cosmological moorings and therefore presumably generalizable to all peoples" (1995:4).

9. Haraway warns that "social constructionism" runs the risk of seeing everything as finally made by people, a view that denies that there is anything at work in the world besides humans and that assumes that humans have complete control over what they construct. This is the flip side of seeing nature as something to be protected or policed into authenticity, a practice that frequently leads to violence against peoples stereotypically identified as close to or identical with nature (Haraway 1991e). Strathern writes that "increasing discourse on the role of 'social' construction in the conjoining of natural and social relations—of the artificiality of human enterprise—has given a different visibility to natural relations. They acquire a new priority or autonomy" (1992a:53). The anthropologist Anna Tsing remarks that "the idiom of the cultural 'making' of nature separates the plastic medium (with its own possible refusals and unpredictabilities) from the cultural template" (1995:115). For similar arguments, see Latour 1987, 1991; Butler 1993.

10. Foucault writes that the earlier tradition of natural history that evolution follows became concerned with classifying living things on the basis of visible structure when a distinctly modern epistemology of representation came into being, one that saw the world ordered by formal relationships. "As Linnaeus [the man responsible for modern taxonomic practice] said, in a passage of capital importance, 'every note should be a product of number, of form, of proportion, of situation' "(Foucault 1966:134). "Hearsay is excluded, that goes without saying; but so are taste and smell, because their lack of certainty and their variability render impossible any analysis into distinct elements that could be universally acceptable" (1966:132). We inherit the biology that grew out of these taxonomic frames as one often concerned with the visible structure of living things; hence the term "life-forms."

11. See, for example, George Marcus's work on dynastic families in the United States (Marcus with Hall 1992), Lutz and Collins's 1993 study of the production of *National Geographic,* and any number of ethnographies of scientists cited in the present book.

CHAPTER 1

1. Not long after I finished my fieldwork, SFI moved to a new site, a 14,000-square-foot building located on thirty-two acres about four miles out of town, up toward the Santa Fe ski basin. The new building is in an affluent residential area inaccessible by public transportation. Public access is blocked by a gate.

2. I write this section in the "ethnographic present," even though SFI has moved. Writing in the past sounded eerie, as if I were writing of a people long dead. At other moments, I do write in the past tense, to recount histories or indicate completed conversations with Artificial Life researchers.

3. "In architecture, Santa Fe Style means adobe, or at least the adobe look, with kiva fireplaces and ceilings of vigas and latillas reminiscent of ceilings in the ancient pueblos of the area. In fashion, Santa Fe Style commonly means turquoise and silver jewelry, full skirts, and velveteen Navajo-style blouses" (Hazen-Hammond 1988:116).

4. This is SFI's trademark, and I am not permitted to reproduce it here.

5. This interpretation was reinforced when I went to the SFI photographic archive and found the photograph from which this framed reproduction was made. On the back, it said

> Dr. and Mrs. Albert Einstein at Hopi House, Grand Canyon, Arizona, 1931 left to right: 1. J-B Duffy, General Passenger Agent AT + SF RR; 2. Herman Schweizer, Manager of Harvey Curios; 3. unidentified Hopi man; 4. Dr. Albert Einstein; 5. Mrs. Albert Einstein; 6. Additional Hopi Indians.

There are four people who fit under number 6.

6. Contrast the naming of computers with how the SFI administration thought of naming buildings on its new campus. In a brochure aimed at potential benefactors, SFI suggested that buildings might be named to honor individual contributors (see Santa Fe Institute 1994a). Hypothetical names presented as examples were "The 'Mary and John Doe' Center," "The 'Jane Smith' Conference Facility," "The 'Mary Brown' President's Office," and "The 'John Smith' Lounge." It should be obvious that these are not generic names but Anglo-American ones. Santa Fe is a town in which half the residents have Spanish surnames; there is not one such name given as a sample, something that points to the ethnic background of the people running the Institute, or at least to the ethnic group that they expect to contribute to their institution.

7. The exceptions during the time I did my fieldwork were the three men who occupied the top administrative positions: president, vice president in charge of finance, and vice president in charge of academic affairs. Since I left, SFI has selected its first woman president as well as its first woman vice president (see Santa Fe Institute 1996, 1997). When vice president for academic affairs Erica Jen was named in October 1996, she expressed a commitment to bringing more minorities and women to SFI.

8. Early forays into looking at science anthropologically, particularly by sociologists of science, called on the tropes of an outdated and colonialist anthropology. Bruno Latour and Steve Woolgar, for example, wrote that scientists might be seen as members of a "tribe" with its own superstitions, myths (the scientific

literature), and ritual practices: "Our anthropological observer is thus confronted with a strange tribe who spend the greatest part of their day coding, marking, altering, correcting, reading, and writing" (1986:49). There are obvious critiques to be leveled at such definitions of cultural life: in treating the lab as a bounded space, sociologists erased continuities between scientific reasoning and various sorts of social reasoning—though, perhaps, precisely the opposite was their goal: Latour and Woolgar say that the anthropological move is "paradoxically . . . intended to dissolve, rather than reaffirm the exoticism with which science is sometimes associated" (1986:29). Early uses of "anthropology" in the sociology of science frequently left out how ideas about agency, gender, economy, and so on intrude on the way problems are framed, ranked, and solved in the lab.

9. Probably least of which is that "the Santa Fe Institute" was the name of an alternative healing arts center before SFI bought the name.

10. This association is also present at Los Alamos, where during one visit, I was escorted to the mobile home office of a researcher, who called it a "frontier cabin."

11. The founding of SFI has much to do with the prestige and clout of physicists in the post–World War II era. Physicists have been known to possess an impulse to colonize and scientize other disciplines, an impulse based in part on the political and empirical success of their physicalist reductionist approaches to understanding the world (see Kevles 1977, Traweek 1988, and Gusterson 1996 on the priestly place of physicists in U.S. political culture).

12. The twentieth century has seen at least two previous attempts at unifying the sciences. In the 1920s, the Vienna Circle of logical positivists sought to construct a neutral observation language that would allow people studying diverse object domains to integrate empirical observations into a unified theory of the natural world. In the 1940s and 1950s, scientists in cybernetics tried to forge a common set of principles for understanding communication and control in organic and machine systems. In both cases, people from a range of disciplines came together to standardize vocabulary and see if there were shared principles governing different objects of study.

13. *Maverick* is a word used again and again to describe this group, and it is a word that evokes Old West cowboy adventuring (see "The Scientific Mavericks of Santa Fe Zero in on Evolution" in the March 16, 1994, edition of the *Japan Times* [Glosserman 1994]). "Maverick" was the name of U.S. settler Samuel A. Maverick, a rancher who did not brand his calves. A maverick is defined by *Webster's Third New International Dictionary* as "an unbranded range animal; *esp* : a calf on the range that is unbranded and not following its mother." The *Oxford English Dictionary* records the colloquial meaning as "a masterless person, one who is roving and casual."

14. *Mondo 2000* promotes a hallucinatory aesthetic, setting hyperactive prose and computer-mutated photo art loose to create dense celebrations of how computer network technologies might allow people to experience ecstatic alternate realities (in the early 1990s, acid guru Timothy Leary was a frequent reference and contributor), of how poststructuralist literary theory might be used

to ironically indulge the bliss of purchasing new technologies, or of how distributed computing may allow people to subvert the agendas of multinationals and governments. *Mondo* continues a tradition very much in line with a U.S. ethos of questioning authority and with making new worlds for ourselves as individuals, and it sometimes edges into an extreme libertarian political position (see Sobchack 1993). Few middle-aged people at SFI were aware of the magazine, even though Artificial Life has been featured and has a celebratory entry in *Mondo 2000's User's Guide to the New Edge* (Rucker, Sirius, and Mu 1992).

CHAPTER 2

1. Most people treated "worlds" and "universes" as synonyms, so I do too. If there were differences in how "world" and "universe" were used, they almost always marked worlds as smaller places, with characteristics that took for granted the physics of a larger universe in which the worlds were nested.

2. Similar views have been expounded by the physicist Ed Fredkin (see Wright 1988).

3. In his dizzying critique of the notion of the mathematical infinite, *Ad Infinitum—The Ghost in Turing's Machine,* Brian Rotman (1993) argues that mathematics—down to the very practice of counting—is enabled by the profoundly theological custom that mathematicians have of imagining themselves abstractly capable of counting to infinity. Rotman maintains that the ideal mathematical subject is a disembodied and omnipotent ghost and that real embodied mathematicians are able to imagine themselves identical to this ghost through waking dreams that they call thought experiments. Reading my scientist's words through Rotman's lens, I would say that if worlds are math and if math is the result of a regulated dream, then worlds are kinds of dreams. Artificial worlds become artificial dreams.

4. The mathematician and physicist Cris Moore reminded me that such machines do not necessarily change the definition of what is "computable." At most, they do some things faster than classical computers.

5. This debate is similar to discussions in Artificial Intelligence over whether a properly programmed computer could ever be said to think. Proponents of "strong AI" consider computers capable of thought, whereas proponents of "weak AI" hold that computers can stand only as models, not instances, of intelligence (see Searle 1980). Some philosophers have defined weak and strong positions for Artificial Life, usually favoring the weak position in their alliances (see Langton 1992; Sober 1992). Others have argued that analogies between Artificial Intelligence and Artificial Life gloss over key differences between the fields (Keeley 1994). Strong Artificial Life, for example, need make no claim for the subjectivity of virtual creatures.

6. Emmeche's title also plays on the literary theorist Leo Marx's *The Machine in the Garden* (1964), a discussion of how technology has both disturbed and remodeled pastoral ideals in U.S. literature and culture.

7. I use this term with reservations. I mean to designate a tradition that includes stories from the Old and New Testament, by which I mean stories of cre-

ation and salvation. I recognize the specificity and multiplicity of Jewish and Christian histories and traditions. What I am trying to index is how biblical themes mix together in the imagination of many atheists and agnostics living in cultures fundamentally shaped by Judaism and Christianity. Using the term "Judeo-Christian" also highlights the hegemony of Christianity in European and American society, drawing attention to the ways Jewish materials have been appropriated into the purposes and narratives of Christian salvation history, which looks for prefigurings of Christ and of the apocalypse everywhere.

8. I thank Kiersten Johnson for remembering this phrase of Aristotle's to me, and for reminding me of how it was incorporated into Judeo-Christian cosmography in the Middle Ages by Aquinas and others.

9. See Risan 1996 for a related discussion of how programmers continually shift the boundaries between self, interface, and simulation in technical practice.

10. The phenomenologist of religion Rudolf Otto has remarked that this definition of miracles is more scientistic than theological. It is exactly the definition to be expected from a science that tries to couch everything in rationalist language: "The traditional theory of the miraculous as the occasional breach in the causal nexus in nature by a Being who himself instituted and must therefore be master of it . . . is itself as massively 'rational' as it is possible to be" (1950:3).

11. The anthropologist Mimi Ito remarked to me that she found it curious that Artificial Life scientists imagined their worlds as closed, as something to watch, not participate in (except as creators). Among the groups of people Ito was working with—people in Multi-User Domains on the Internet—there was a sense that computers and humans were always both participating; it was an open system, not one in which "the computer world" was seen as coming to be independent of humans (see Ito 1997).

12. The idea that there might be life on other planets is hardly exclusive to twentieth-century science fiction, though this is where it finds its recent commonsense moorings. Debates about extraterrestrial life reach into antiquity, as Steven J. Dick argues in *Plurality of Worlds: The Origins of the Extraterrestrial Life Debate from Democritus to Kant* (1982). In the eighteenth century, these debates became connected to theological debates about whether God, being capable of all things, had filled the universe with all possible worlds. For more on this, see Michael J. Crowe's *The Extraterrestrial Life Debate 1750–1900: The Idea of a Plurality of Worlds from Kant to Lowell* (1986).

13. I might highlight my own use of Le Guin, an author whose storytelling is highly informed by cultural anthropology. Le Guin was raised by the prominent anthropologists Alfred and Theodora Kroeber.

14. Thanks to Julian Bleecker for setting me on the trail of colonial tropes in Artificial Life. See Sardar 1996 and Healy 1997 for treatments of frontier and colonial imagery in cyberspace more generally.

15. The colonial themes of Artificial Life research are nowhere so obvious as in plans for producing self-replicating robots that can live in outer space. These plans, now somewhat forgotten, originated in early 1980s research done by NASA to investigate the possibility of sending autonomous robots to the moon.

The idea was to get robots to extract lunar resources for use by the United States and its allies. Rodney Brooks of MIT was part of this project, and has more recently been constructing robots with Martian exploration in mind.

The idea of robots as a worker class is not new; indeed, the word *robot* itself is derived from the 1921 Czech word *robota,* which means "forced labor" (see Channell 1991:139). The word originates in a Czech play called *R.U.R.* by Karel Capek, in which robots are created to supplant human factory workers. Ultimately, in an allegory for proletarian uprising, they rebel, taking over the means of production—and reproduction, when they learn how to procreate.

16. Listening to the "American" resonances of some Artificial Life rhetorics, I suggest that the creation and frontier imagery often used to motivate Artificial Life research at Santa Fe has something in common with that deployed by a somewhat different set of people: the Mormons. The Mormon church, whose founding members set out to create a New Jerusalem in the American West, holds that in the afterlife, when Jesus has returned, married Mormon couples who have practiced Christian charity will inherit new worlds to populate. Like Adams and Eves posted on the final frontier, they will begin anew. Artificial Life researchers might be seen as enacting with their computers what Mormons hope to achieve in the world beyond.

17. As if to confirm the psychedelic connection, the famously hallucinogenic 1960s band the Grateful Dead have taken an interest in fertilizing the gardens of Artificial Life. The slow-trucking backup band for the late Jerry Garcia has decided to put some digital organisms on their tab, sprinkling $10K into Ray's multik project to extend Tierra into the Internet.

18. George Kampis makes a similar argument from within the philosophy of Artificial Life in "The Inside and Outside Views of Life" (1995).

19. Note that while telescopes and microscopes transmit, amplify, and create visual data on real phenomena, computers manufacture in very different ways the phenomena they allow us to see.

In her study of high energy physicists, the anthropologist Sharon Traweek writes of watching the traces of subatomic particles in a bubble chamber, and she notes that the visualization technology that makes these traces available produces the sense that one is directly watching events unfold in the subatomic world. The bubble chamber, while it has some claim to represent a reality beyond machines, is similar to the computer in that it does a good deal of manufacturing of the very realities it attempts to disclose: "It was not easy to remember that I was looking at signals produced by a machine designed to react in a stylized way to the debris from a collision occurring under highly controlled and artificial conditions. I was not looking through a window on the world of subatomic particles; I was doing something more like reading dinosaur tracks. It was the 'real-time' activity in the chamber that created the impression of seeing" (Traweek 1988:54).

20. The notion of evolution as a "tape" is discussed in the evolutionary biologist Stephen Jay Gould's *Wonderful Life* (1989). Artificial Life scientists hope to make Gould's metaphor do literal work but fail to demonstrate how "rewind-

ing" simulated evolutionary history can make up in a straightforward way for any " 'lack' of reality" in computer models.

CHAPTER 3

1. Most simulations offer models of animal rather than plant behavior. For foundational work in modeling plant growth, see Lindenmayer and Prusinkiewicz 1989.

2. As I wrote this chapter, many programs I discuss became available over the Internet. The reader with access to the World Wide Web may be interested to search for sites offering details and downloads of Tierra, PolyWorld, and Echo. While hardly essential for reading this chapter, this sort of web walking would complement the tour I lead here.

3. The technical reader should note that Tierra is instantiated in a "virtual computer," a computer that exists as a simulation within a real computer. When Ray began thinking about creating self-replicating programs, he was advised not to write code that would replicate using the machine codes or operating systems of real computers. There was a concern that creatures might "escape" from their home machine and wreak havoc on other machines. The machine code in which Tierrans are written works only in the virtual Tierran computer; outside, Tierrans exist as data, not programs.

4. The similarity between the way bases in DNA pair up and the way os and 1s pair up in the Tierran "genetic code" is highly metaphorical. In DNA, bases pair via hydrogen bonds. Zeros and 1s in Tierra pair up because Ray decided on this convention.

5. In 1993, Ray collaborated with the animators Tim Wilson and Thomas Hollier of the Anti-Gravity Workshop and the filmmaker Linda Feferman to produce a video vision of Tierra. When one watches this video, new layers of rhetorical apparatus are mobilized to make Tierra an alternative world. Situated in the midst of a *Star Trek*-like set, Ray narrates us into a world shimmering beyond his computer screen. Our gaze, commandeered by the camera's eye, passes through the cathode ray looking-glass and is catapulted into a dreamlike fantasy flight over the rich "ecosystems" of Tierra. Using a style of narration similar to that employed in nature programs, Ray guides us over landscapes of digital organisms being born, competing, and perishing (see Santa Fe Institute 1993).

6. The medieval alchemist Paracelsus (1493–1541) literalized the seed and soil metaphor in a bizarre and primordial artificial life recipe:

> Let the semen of a man putrefy by itself in a cucurbite (gourd glass) with the highest putrefaction of the *venter equinus* (horse dung) for forty days, or until it begins at last to live, move, and be agitated, which can be easily seen. After this time, it will be in some degree like a human being, but nevertheless, transparent and without body. If now, after this, it be every day nourished and fed cautiously and prudently with the arcanum of human blood, and kept for forty days in the perpetual and equal heat of a *venter equinus,* it becomes, thenceforth a true and living infant, having all the members of a child that is born from woman, but much smaller. This we call a homunculus; and it should be afterwards educated with the greatest care and zeal, until it grows and begins to display intelligence. (Quoted in Cohen 1966:43–44)

7. Here, Hayles is reading the biologist Richard Dawkins's account of his interaction with his program, Blind Watchmaker, which I discuss later in this chapter.

8. Some biologists argue that gendered notions of information and material, seed and soil, blocked the recognition of the role of the ovum in conception (until 1826) and contributed to views of nucleic acid as active and cytoplasm as passive (see Fausto-Sterling 1992). It is still true today that in stories about mammalian conception, the sperm is often said to "activate" a passive egg and set its cytoplasmic machinery into motion (see Martin 1991). The active role of the egg's cytoplasm in regulating the DNA of both egg and sperm is overlooked.

9. This focus on form inhibits early ruminations about automata reproduction. Von Neumann, in *Theory of Self-Reproducing Automata* (1966), was intrigued by the possibility that life and reproduction might be formalized. As Arthur Burks, editor of this posthumously published volume, put it, "He wished to abstract from the natural self-reproduction problem its logical form" (quoted in Langton 1989:13). On the topic of how Artificial Life creation tales often repeat Bible stories, compare the way von Neumann's creation tale mirrors that offered in Genesis:

> The constructing automaton floats on a surface, surrounded by an unlimited supply of parts. The constructing automaton contains in its memory a description of the automaton to be constructed. Operating under the direction of the description, it picks up the parts it needs and assembles them into the desired automaton. (von Neumann 1966:82)

> And the earth was without form, and void; and darkness was on the face of the deep. And the Spirit of God moved upon the face of the deep. . . . And God said, Let us make man in our image, after our likeness. (Gen. 1:2–26 in the Bible, King James Version)

10. Even with the "feminine" imagery in Tierra, the masculine meanings sewn into "seed" are difficult to unstitch. If we are eager to exile these meanings, we might engage in a Derridian reading, in which the figure of "seed" renders origins problematic, since it begs the question of where potential comes from. We could also listen to readings of "matter" that invest it with generative principles. Judith Butler writes,

> The classical configuration of matter as a site of *generation* and *origination* becomes especially significant when the account of what an object is and means requires recourse to its originating principle. When not explicitly associated with reproduction, matter is generalized as a principle of origination and causality. In Greek, *hyle* is the wood or timber out of which various cultural constructions are made, but also a principle of origin, development, and teleology which is at once causal and explanatory. . . . *Materia* in Latin denotes the stuff out of which things are made, not only the timber for houses and ships but whatever serves as nourishment for infants: nutrients that act as extensions of the mother's body. Insofar as matter appears in these cases to be invested with a certain capacity to originate and to compose that for which it also supplies the principle of intelligibility, then matter is clearly defined by a certain power of creation and rationality that is for the most part divested from the more modern empirical deployments of the term. (1993:31–32)

While these alternative meanings of matter are available, they are mostly muted in Artificial Life.

11. What allows people to see life emerge in Tierra is a systematic forgetting of the metaphors that make such seeing possible (see Doyle 1997b:127–128). Researchers take computers as blank slates that come to them from nowhere, ready to be impressed with formal programs that will turn raw silicon material into worlds and life. What is effaced are the conditions under which computers are produced. The hardware and software entrepreneurs of Silicon Valley, the usual heroes of stories about the manufacture of computers, work with machines often put together by undercompensated third world women who assemble computers in home-working situations (see Ong 1987). In Artificial Life, the only allusion to the mostly female labor that puts together the raw material for simulations comes in the form of Ray's "mothers" and "daughters" laboring in the computational colony of Tierra. Interestingly, the word *computer* once referred to people who, working in factorylike settings, calculated fragments of large astronomical problems (see Schaffer 1994) or later, artillery firing tables (see Keller 1995). By the mid-twentieth century, most of these people were women. Might the diminutive computers in Tierra represent a return of this repressed?

12. The genebank, from which frozen codes come to life, is a kind of magical realist version of GenBank, the Human Genome database maintained at Los Alamos.

13. *Webster's Third New International Dictionary* defines *to father* as "beget" or "to be the founder, creator, or author of." *Mother* has no such meanings, defined simply as "to give birth to" or "to give rise to," emphasizing the passivity of this role (and see Franklin 1995, 1997).

14. According to Tom Ray, of course, Tierran organisms *are* real.

15. Thanks to Kiersten Johnson for this turn of phrase.

16. Following this logic, computers are not "artificial worlds" but "artificial wombs." In this connection, I recall that at one conference I attended, roboticists demonstrated machines that looked autonomous but that were attached through a cable to a computer running a motion-governing program. The cable was referred to as an "umbilical cord," suggesting the computer as a mother providing a life force.

17. There are historical instabilities in the assignment of the Virgin Mary to a position subordinate to God. Some are inventoried by the linguist and psychoanalyst Julia Kristeva in "Stabat Mater" (1983), which discusses how Mary was imagined at some moments as a nearly divine being whose life prefigured and paralleled Jesus' and who, as mediator between God and humanity, might be identified with the Holy Ghost. In more familiar tales, Mary became a wholly human mother who supplicated herself at the feet of God (simultaneously her son and father). If computers are feminized virgin media for the transmission of form, this does not mean they cannot also be symbolically morphed into partners or even originating principles in the animation of artificial life (see note 10 on matter, above).

18. The description of DNA as a text that is translated and transcribed secularizes the Judeo-Christian idea that the Word determines the form of matter in Creation. Genetic determinism replaces a doctrine of predestination when the divine and unquestionable authority of the biblical text is replaced by the

unquestionable authority of the genetic text. "It is genetic" stands in for "It is written."

The filmmaker Dmitry Portnoy reminded me that identifications of textuality with vitality were discussed by Jewish mystics who thought of the Torah as a living thing. He also noted that medieval priests who spent their days illuminating manuscripts thought of the reproduction of texts as the transmission of the vital Word, for which they were both authorized masculine messengers and feminized vessels. We might connect recent fusions of text and life not just to histories of code metaphors in genetics but also to understandings of text as vital in particular mystical traditions. Indeed, we might see the story of creation in the Gospel According to John as a template for Artificial Life cosmogonies:

> In the beginning was the Word:
> the Word was with God
> and the Word was God.
> He was with God in the beginning.
> Through him all things came into being,
> not one thing came into being except through him.
> What has come into being in him was life.
> John 1:1–4 in the New Jerusalem Bible

19. There is another, less frequently noted, implosion here: the collapse of two different meanings of information. The first, defined by the mathematician Claude Shannon, is simply a quantitative measure of the complexity of a linear code or message and has nothing to do with what the code or message means. The second, associated with computer programming, attaches to the concept of instruction or program, for which meaning is of utmost concern (see Keller 1995). The fusion of these meanings in "genetic information" allows genes to become "messages" that give "instructions" to the organism.

20. In late-twentieth-century pop culture, this idea informs Cartesian cowboy stories like William Gibson's *Neuromancer* (1984), in which the body, known as "the meat," is left behind for the formal gridlike universe of cyberspace. In Gibson's book, "the meat" is associated with things sexual and female, like the Molly character, whose seductive presence threatens to drag the main male character, Case, into the unruly corporeal world.

21. This similarity between gene and mind is enabled by what Evelyn Fox Keller has called the discourse of gene action, which has

> functioned historically to endow the material gene simultaneously with properties of life and mind. As a unit of transmission, the *gene* was credited with permanence; as an autonomous entity capable of reproducing itself, it was credited with vitality; as ontologically prior to life, with primacy; as the locus of action, with agency; and as capable of directing or controlling development, with mentality. Part physicist's atom and part Platonic soul, the *gene* was assumed capable simultaneously of animating the organism, of directing, and of enacting its construction. (1996:4)

22. Computer viruses were formally defined in 1983 by Fred Cohen, who put together a simple example to demonstrate to his University of Southern California computer security seminar. The first notable infection of a system happened in 1986 with the Brain virus (Levy 1992b).

23. Computer viruses are familiar enough that the popular press often refers to artificial life as viral: an article about Tierra in *Newsweek* tells us that the notion of "electronic biodiversity led Ray to propose seeding the Internet with progenitor viruses" (Rogers 1995:65). Broader discussions of computer viruses capitalize on the biological metaphor and describe the need for individual or networked protection in terms borrowed from immunology, in terms that invoke images of bodies as self-contained entities that must be protected from outside threat. These discussions import from popular and medical discourse anxieties about sexual contamination in populations, and sometimes proffer "safe sex" tips for computer use. Discussions of how networks might be protected also employ language reminiscent of that used to describe nation-states under threat from without and within. Viruses are described using the language of foreignness, illegality, and otherness. Newer figurations of the immune system as a "flexible" and "adaptive" system are also beginning to infiltrate computer talk (see Haraway 1991c; Martin 1994).

24. In characterizations of the "primitive" in late-nineteenth-century and early-twentieth-century human evolutionary biology, white male scientists often drew analogies between (white) "women" and members of "lower races," locating both as falling short of the more perfect and advanced forms exemplified by white men (see Stepan 1993). While no one in Artificial Life would intentionally replicate this sort of reasoning, the ghosts of its metaphorical structure and accompanying imagery still haunt their talk.

25. Returning to the issue of how artificial life creatures come alive when possessed of an "endearing quality," I recall a famous children's story entitled *The Velveteen Rabbit*, in which a stuffed rabbit doll becomes "real" when a little boy cares for it. On this model, artificial life creatures are lovers/toys, a reading in tension with constructions of such creatures as independent things for which creators must not care. At moments of emotional attachment to their creations, Artificial Life researchers share something in common with the zoophiles known as "furries," people who have sensual attractions to anthropomorphic animal art (which may range from Disney cartoons to science fiction erotica featuring human/animal hybrids). I am inspired to this connection by "Nooga's Artificial Life Page," a website run, so the copy would have us believe, by a cat named Nooga. Nooga, a kind of hypercute incarnation of Alice in Wonderland's companion, "Kitty," guides us on a trip through the looking-glass, into an Artificial Life program called Creatures, populated by adorable cartoon animals that mate by kissing.

26. This cyborg of color is emphatically not the kind of active and transgressive agent championed by Donna Haraway in her socialist-feminist writings (see Haraway 1991d).

27. Other, similar, programming paradigms were invented independently at around the same time, including evolutionary programming and evolution strategies (see Goldberg 1989a and Mitchell 1996 for histories).

28. The genetic algorithm is usually used to solve problems for which the answer is a nonobvious choice out of an enormous set of possibilities. For this

reason, populations contain only a fraction of candidate solutions to a given problem.

29. There are a variety of selection and mating schemes, but fitness proportionate reproduction with subsequent crossover has been most popular. For information about other selection and mating regimes (e.g., tournament selection, rank selection, and elitist selection), see Goldberg 1989a and any International Genetic Algorithm conference proceedings.

30. "Crossover" is not quite synonymous with the sexual recombination that the process of swapping code is meant to recall. In the genetics of heredity in sexually reproducing diploid organisms (which genetic algorithms take as inspiration if not literal model), "crossover" refers to the trading of material between homologous chromosomes during meiosis (i.e., within the gametic cells of one organism). It is a kind of recombination but is not identical to sexual recombination—a point Stephanie Forrest clarifies (1993:873). One could argue that because each genetic algorithm parent is made of one chromosome and so is "haploid," the conceptual difference between crossover and recombination is unimportant.

31. To understand how a genetic algorithm wends its way through a problem, consider the following elementary example from Forrest 1993: Suppose we want to find a maximum value for the function $f(x, y) = yx^2 - x^4$, where x and y are integers and can vary between 0 and 7 (this function is solvable analytically, but if it were not, it would be a perfect candidate for solution by the genetic algorithm). Possible values for x and y can be represented in a bit string by making sublengths of the string represent each parameter. Since x and y vary between 0 and 7, we only need three bits to represent each number in binary (000 being 0 and 111 being 7), meaning that we need a string only six bits long to represent any possible solution. A solution where x is 1 and y is 5 would be represented as 001101, where the first three digits code for $x = 1$ and the second for $y = 5$. The fitness of a bit string like this can be ascertained by decoding the bit string into decimal values and plugging these into the function $f(x, y) = yx^2 - x^4$. In a genetic algorithm program written in Pascal, the work of determining the fitness might happen in a subroutine looking something like the one in figure 23.

32. This effectiveness is always relative to a set of expectations and criteria. Indeed, some people debate how well genetic algorithms can work without persistent shepherding. In what follows, I do not argue that the cultural ideas constitutive of genetic algorithms render them operationally incoherent. They still "work," an outcome that does not necessarily mean that they do so because they twin nature.

33. We might note, following the sociologist Howard Kaye (1986), that understanding natural selection as an agent is to read our purposes onto nature; natural selection is not an agent and cannot in fact even be a process. It is a "statistical artifact, not a set of operations and actions organized and directed towards some end" (Kaye 1986:53).

34. This progressivist view of evolution is recent. Prior to the Enlightenment, most narratives of nature had human beings degenerating, continuing a descent from perfection that began with expulsion from Eden. With the rise of Protes-

```
function        objfunc (x:real, y:real): real;

{fitness function f(x,y) = y times x squared - x to the fourth}

const

begin

        objfunc := y*power(x,2) -power(x,4);
        {power(a,b) returns a raised to the b}
end;
```

Figure 23.   Subroutine that determines fitness, from a genetic algorithm program written in Pascal.

tantism and Enlightenment beliefs about progress, religious stories of human history began to be told as stories of salvation achieved through hard work. Evolutionary biology secularized these tales.

35. To have a sequestered germ-line means that changes to somatic cells do not affect the germ cells involved in the transmission of genetic material.

36. This "architecture" consists of a "neural net," a network of processing elements (thought of as artificial neurons), connected by lines that transmit signals between elements. Networks receive inputs that activate elements in ways dependent on their connections. Networks respond to inputs representing different problems and become better at solving problems both as a result of the evolution of their architecture (wiring diagram) and, potentially, as a result of a kind of unsupervised, associative learning (by adjusting connectivity "weights" between elements). Some genetic algorithmists have tried to evolve neural nets by encoding connection weights in bit strings (see, e.g., Harp, Samad, and Guha 1989; Miller, Todd, and Hegde 1989; Gruau 1993). In these systems, as in PolyWorld, initial weights are inherited from parents in Darwinian fashion; changes to connectivities acquired in organisms' "lifetimes" are not transmitted.

37. This statement encodes masculinist scientific impulses to appropriate the active female role in procreation and is part of the same tradition that brought us obstetrics and artificial wombs. The out-of-body experience to which it refers requires a masculine identification with one's "seed." Such identification is serviced by dominant definitions of sex and sexual pleasure after a masculine model that fixes ejaculation as the defining moment of sexual acts (see Irigaray 1977). Note also that this view reduces the value of reproduction to its production value in making offspring. Thanks to Heather Paxson for this observation.

38. In evolutionary psychologists Geoffrey Miller and Peter Todd's models of sexual selection, for example, bit strings are endowed with genetically encoded "mate preferences" and assigned to explicit sex roles in crossover: "To create the next generation, a 'mom' individual is first picked from the old population. . . . Since it takes two to tango, a 'dad' is next selected from the population. . . . (Of course, individuals do not have a sex per se; how each individual is picked determines whether it plays the mom or dad role in each mating)" (Todd and Miller

1991:551). Based on her "mate preferences," the "mom" may refuse the "dad" and try again. "If a mom cannot find a suitable mate after five times through the population, she is deemed hopelessly picky, and a new mom is chosen" (Todd and Miller 1991:551). Though Todd and Miller claim that sexes in their model are fictive (as they put it in a later article, "these are temporary gender roles, not genetically determined sexes" [Miller and Todd 1993:24]), they call on visions of women as coy and men as promiscuous, visions embedded in sexual selection theory since Darwin, which in the present day rely on a sociobiological rationale that roots these dispositions in the logic of genetic and parental investment. (The idea is that males, in many species, invest proportionally less in each of their gametes [sperm] than do females, who not only produce fewer gametes [eggs] but must also stay with "fertilized" eggs through gestation. Sociobiologists claim that this leads to differences in "genetic strategies" and dispositions between the sexes; males are promiscuous and females picky. In order to believe this, one must first accept the metaphysical conceit that genes are selfish entities that deterministically construct organisms.) Miller and Todd's appeal to "gender roles" does nothing to qualify the sexual stereotypes built into their model. Indeed, the very idea that there are two genders makes it impossible to talk about without invoking "natural" sex difference: "The presumption of a binary gender system implicitly retains the belief in a mimetic relation of gender to sex whereby gender mirrors sex or is otherwise restricted by it" (Butler 1989:6). Todd and Miller link gender roles tightly to reproduction, completely entangling sex, gender, and desire.

39. Of course, such simplification can be easily remedied, as Goldberg (1989a) suggests when he proposes a diploid model in which bit strings "fertilize" one another. Such modifications, though, do nothing to banish determinist views of sex difference as genetic.

40. The anthropologist Heather Paxson (personal communication) cautions that the category of "sex" as sexual difference can exist without normative heterosexual society, and offers classical Athens as one example. Difference can be recognized without entailing heterosexuality.

41. This conception ordered much of early anthropology. European and U.S. anthropologists often spent their time searching for biological "aunts" and "uncles" in societies that turned out to organize kinship around very different terms and relationships (see Fox 1967 for an example of this traditional approach).

42. For a recent account of changing biological rhetoric about race, see Haraway 1997.

43. The biogenetic grid was also used to prevent intermarriage between whites and Chinese-Americans, Japanese-Americans, and Filipino-Americans, while Mexican-Americans, Latinos, and Asian Indians were considered white at some times and not others (see Frankenberg 1993; Haney López 1996). The grid was used inconsistently and contradictorily where the category of African-American was at issue; in this case, many U.S. states designated someone as black using the "one-drop" rule, which held that fractions of black ancestry down to 1/32 made a person black. This rule has origins in maintaining the category of slaves as property and in preventing people from voting. Bilateral inheritance is not erased here; rather, it is given a quite different meaning.

44. Note that these laws are resisted, contested, and appropriated strategically by Native Americans.

45. For work that attempts to retrieve nonwhite histories in the sciences of chaos and complexity, see Eglash 1995, forthcoming. For general treatments of the racialization of science, see Harding 1993.

46. There are two versions of Echo, one written by John Holland in BASIC and one by Terry Jones in C (with Stephanie Forrest serving an advisory role). The second Echo, known as "SFI Echo," is described here. Since my fieldwork, Holland has proposed a variety of extensions to Echo models (see Holland 1995).

47. So fundamental is competition to neo-Darwinism that even cooperative, symbiotic, or mutualistic relations are understood as selfish and competitive. Competition is even extended to include situations "where there is no interaction at all, where 'competition' denotes an operation of *comparison* between organisms (or species) that requires no juxtaposition in nature, only in the biologist's own mind. This extension, where 'competition' can cover all possible circumstances of relative viability and reproductivity, brings with it, then, the tendency to equate competition with natural selection itself" (Keller 1992c:125).

48. In mathematical ecology, this is especially so in the use of Lotka-Volterra equations, which describe competition between two species. Developed independently by the physicists A. J. Lotka and V. Volterra in the 1920s, they have become canonical in evolutionary biology (see McIntosh 1992). Holland (1994:331) claims that Echo is a good model of evolutionary dynamics in part because it produces results consistent with these equations.

49. One scientist gave me a copy of the "Career Guide for Women in Science" (1988) drafted by a group of women faculty and professionals at Cornell University. This document recommended that women deemphasize differences from men. Some excerpts: "Any excess use of cosmetics or of fragrance can generate the wrong images. While men leave framed pictures of the family on their desks, women scientists in the physical sciences usually do not. Such reminders that women have families and/or children conjure up images of motherhood and mothering; not leadership and business efficiency" (50). "To . . . fellows at work, the inflated image of pregnancy can be treated as a temporary condition, and any necessary maternity leave should be treated in the same way as one would treat a gall-bladder operation. The female aspects of the condition should not be emphasized" (45). "Some women professionals with very competitive jobs prefer to keep their private and public lives completely distinct, separating friends they can rely on from colleagues who are competing for the same contracts. This is often the case for very successful women in science" (46). "Many successful women (and men) scientists . . . dress simply so that they blend in best with their peers, and . . . have their hair cut in simple styles which require little care" (47).

50. These generalizations are not meant to include *all* women, at all times, in all contexts. Some women, enculturated outside the tradition of U.S. science, may feel less pressure to adopt masculine codes of dress or behavior, and this has to do with the ways feminism has developed in different parts of the world. In Anglo-American contexts, liberal feminism has been premised on attempts to enfranchise women on terms set by male norms. As women have modeled themselves after

the masculine subject of liberal humanism, it has been difficult to validate difference without validating sexist readings of that difference. Women socialized outside this context may not always feel bound to emulate male models. There is evidence that there are more women in science in countries such as India, Spain, and Mexico partly because class is a more important determinant of participation in elite jobs than is gender. For discussions of how women scientists (both in the U.S. context and internationally) experience the world of science, see articles in *Scientific American* (Holloway 1993) and *Science* (Barinaga 1994).

51. Many thanks to Cris Moore for early conversations on the science and politics of SFI economics.

52. The economist Michael Rothschild, author of *Bionomics: The Inevitability of Capitalism* (1990), believes this strongly. Rothschild is president of the Bionomics Institute, dedicated to promoting the identity between economies and ecologies. The link to Artificial Life is Tom Ray, who is affiliated with this institute.

53. Although there is a tight relationship between the social positions of SFI scientists and their economic theories and concerns, it is not deterministic. People in underprivileged positions might well find SFI science compelling: capitalism works through instilling in people of many positions the sense that they can become rich if only they work hard enough. And wealthy people can well align themselves with radical causes and recognize the structures that make them wealthy at the expense of others—and can do this even as they perpetuate practices (like investment) they find problematic. What makes most SFI scientists' visions of the economy as a self-regulating entity so leashed to their social position, however, is the fact that they do not have social or economic experiences that would contradict or complicate their models, that would render apparent how the highly mathematical character of high finance presupposes particular educational experiences, how wage structures often depend on gender inequality, how one's opportunities to rent or buy property may depend on whether lesbian and gay domestic partners are recognized as legal entities, and so on. Not only do researchers' experiences not contradict their models, they give them a closer understanding of the very economic phenomena in which they are interested.

CHAPTER 4

1. I do not mean to suggest that everyone in Artificial Life is an atheist. I have not done an exhaustive survey. Note also that being religious, in a traditional sense, is hardly at odds with being a scientist. A great many scientists working in disciplines from astrophysics to molecular biology maintain committed religious belief.

2. The title of this book is itself a play on Eugen Herrigel's 1948 *Zen in the Art of Archery*, which teaches how to achieve unity with the world through a practice of archery in which the archer and target are experienced as one.

3. At many moments, the use of Zen as a frame for vision simply restated an empiricist picture of the world, seeing things as they "really" were. But let me be clear that I do not think that any use of Zen in Artificial Life or in scientific

work in general will necessarily rehabilitate objectivism. There are other ways to bring traditions of meditative mysticism and science together. Francisco Varela, a prominent Artificial Life figure in Europe, has been a practicing Tibetan Buddhist for quite some time and has coauthored a book, *The Embodied Mind,* about how cognitive science might learn from Buddhist meditative practice (see Varela, Thompson, and Rosch 1991). The book argues for a science that never assumes it can know the world apart from our ever-changing experience of it. Though I cannot explore them here, the differences from Zen-inflected epistemologies of computational Artificial Life are both subtle and great.

4. The rhetorician Kiersten Johnson pointed out to me that Langton's resurrection story repeats a well-known narrative in Middle Eastern religion and myth, a narrative about a young god who must be torn apart and reassembled before he can become fully divine. Osiris and Jesus are examples of such gods.

5. Recall that one Europe-based scientist felt that the popular press presented Artificial Life as springing from "Zeus' head in Santa Fe, New Mexico." Giving a mythic spin to Langton's mystical experience, we might see Langton as that Zeus.

6. In one interview I conducted with Langton, he discussed the quasi-religious lessons he had learned from his crash—and here I slide from analysis of texts about Langton to ethnographic interview data. Before his accident, Langton took the world as it was given to him. Afterward and during his convalescence, he radically rethought his relationship to reality. He was no longer an athletic young man but rather a bedridden and temporarily disabled person. He had to discard his old identity and fabricate a new one that took into account his physical limitations. Langton's process of self-reevaluation led him to see the world as full of alternative possibilities, both for constructing personal identity and for understanding reality itself. As Langton tells the story, it is about awakening, but also about his confrontation with the absurd in human existence. While we talked, he sketched the grid in figure 24 and explained that there were four ways of dealing with this absurdity. Langton said that few people had written convincingly about embracing alternative A. One person who had, however, was Kurt Vonnegut, in *Cat's Cradle* (1963). In this book, the doctrine that one could recognize and live with the fundamental absurdity of the cosmos was embedded in the fictional religion of Bokononism. According to Bokononism, all truths are lies, and life should be led believing in those packs of lies that allow us to act effectively in the world and that make us happy. Langton mused that if he had to ally himself with any one religion it would be Bokononism.

	Accept absurdity	Reject absurdity
continue living and working	A	B
give up	C	D

Figure 24.   Langton's view of four ways of dealing with the absurd in human existence.

Langton's avowal is fascinating for a few reasons. First, it expresses a commitment to a theory of reality as socially constructed. Second, it expresses allegiance to a religion that is really a nonreligion (Bokononism resembles a more ironic and perhaps cynical version of Zen, as conceived in the West). Third, it follows a definition of religion that is almost exactly the sort of general and agnostic definition promulgated by secular anthropology—religion as an individual, optional, perspective on the big questions. That Langton finds it appealing is not surprising, given his anthropological background. Interestingly, Vonnegut also did work in anthropology, and has remarked on occasion that anthropology has served him as a surrogate religion. Vonnegut's M.A. thesis was not accepted by the department at the University of Chicago, but some twenty-five years later, when he had become famous, they reopened his files and granted him a degree belatedly for one of his published works, which, it turned out, was *Cat's Cradle*. Langton explicitly tied his appreciation for Vonnegut's Bokononism to anthropology. He referred me to the anthropologist Roy Rappaport's *Pigs for the Ancestors* (1967), in which Rappaport argued that the Tsembaga Maring of New Guinea maintained local ecological balance through a practice of sacrificing pigs to ancestors, a practice that never needed to refer to the "scientific" facts of ecology to be effective. Here, Langton contended, were a set of effective lies or constructed truths. Artificial Life, Langton opined, may trade on true or false epistemological premises—just like any science—but the measure of its validity should be whether it produces explanations adequate to our purposes of prediction.

7. The gender codings of researchers, evolution, and computers can shift during these moments. Nature or evolution, often personified as feminine and embodied in the passive computer, becomes identified with a masculine god, and Artificial Life researchers' minds become feminized spaces waiting to be inseminated with the informatic logics of evolution. At some moments, normatively male researchers are described as "midwives" rather than "fathers" of artificial life (see Levy 1992a), those agents (frequently female) who assist in bringing forth new life but who do not act directly as agents of its production. The Artificial Life researcher assumes a supporting, not generative, role in the creation of life. The generative power is given over to a deified masculine evolution, and Artificial Life work is not only justified but also becomes something of a sacred calling. Keller's (1985) analyses of the writings of Francis Bacon can shed some light here: Keller notes that in the Baconian scientific heritage, scientists are not just in the business of dominating nature but are also nature's servants. The goal of science is not to violate but to master nature by "following the dictates of the truly natural" (36). A complex dialectic emerges: "Science is to be aggressive yet responsive, powerful yet benign, masterful yet subservient, shrewd yet innocent, 'as if the divine nature enjoyed the kindly innocence in such hide-and-seek, hiding only in order to be found, and with characteristic indulgence desired the human mind to join Him in this sport' [Bacon's words]" (37). Nature, previously coded feminine, becomes masculine as it becomes divine, while the scientific mind becomes feminine and receptive. "To receive God's truth, the mind must be pure and clean, submissive and open. Only then can it give birth to a masculine and virile science. . . . Cleansed of contamination, the mind can be impregnated by

God and, in that act, virilized: made potent and capable of generating virile off-spring in its union with Nature" (38). "The scientist himself has assumed the procreative function that Bacon reserved for God: his mind is now a single en-tity, both phallus and womb" (42).

8. This was a story that a number of people told me, in various forms. One computer science professor saw humans left behind while the robot heirs to evo-lution sallied into the extraterrestrial future:

> As to robotic life-forms, they do not yet exist, but I firmly believe that they are the next great phase of matter in the universe and will ultimately conquer biological life-forms. Large populations of replicating robotic factories can reproduce both by fac-tory reproduction and by splitting (i.e., when a new factory is built). Robots would be able to survive space travel, store backup copies, etc. The advantages over organic life-forms are too great. Hopefully robotic life-forms will allow organic life-forms to continue to exist on the surface of our planet.

CHAPTER 5

1. A recent summary of the Gaia hypothesis argues that

> the global regulation of geophysical values over evolutionary time may be likened to the regulation of the body temperature of a mammal over a period of decades; this allows one to speak meaningfully of a Gaian "physiology," of the surface of the Earth as being alive. . . . [T]he atmosphere, far from being a sterile container for life, is inseparable from it, like the shell of a tortoise or the nest of a bird. The atmo-sphere is at once life's circulatory system and its skin. . . . The oceans and air of Earth appear to be continuously *physiologically* stabilized, as are the body chemis-try, internal temperature, salinity and alkalinity of many organisms. (Sagan 1992:371–373)

2. As the philosopher Marc Lange suggests in "Life, 'Artificial Life,' and Sci-entific Explanation" (1996), socially and historically specific intuitions about life influence the sorts of entities scientists include in the realm of the vital. In antiquity and up through Galileo's time, some natural philosophers saw the mo-tion of planets as a sign that they were alive. Opponents of this view argued that planets do not ingest, grow, or reproduce, as animals and plants do. Advocates of planetary vitality replied that these activities were, like celestial motion, sim-ply instances of self-motion. Explanations of celestial motion in terms of grav-ity finally settled the matter. In another example, debates about whether fungi were alive were resolved when scientists began to assume their vitality a priori and search for signs of growth and reproduction that had gone previously un-noticed (Lange 1996).

3. The "form" that Goodwin writes of is that of entire organisms. It should not be confused with the abstract form that many Artificial Life researchers speak of and often compress into the informatic aspect of the hereditary material.

4. In *A Feeling for the Organism* (1983), Evelyn Fox Keller details the life of the geneticist Barbara McClintock and explains McClintock's distinctive way of studying maize genetics. Instead of standing apart from her genetic samples, McClintock imagined herself in the cytoplasm, seeing things from the corn genome's vantage point. During the 1950s and 1960s, when molecular biology

was making grand claims about having found the secret of life in DNA, McClintock questioned the field's central dogma (that DNA makes RNA makes protein). Her work on genetic transposition upset the linear theories of molecular biology and later became important in understanding genetic regulation. Keller explains why it took so many years before McClintock's innovative work was recognized. McClintock favored a nonlinear, noninformatic account of genetics at a time when such explanations were not popular. In addition, she was a woman in a male-dominated scientific field and employed modes of explanation emphasizing the subjective and participative. These ways of knowing were hardly logically connected to her being a woman but were symbolically associated with the feminine in the eyes of a masculine scientific community. Keller argues that science would do well to erase the stigma attached to these ways of knowing, to see that they are not only or always available to women and that they can represent valuable resources for scientific inquiry.

5. Roboticists might be happier with a definition of life offered more than three hundred years ago, by the political philosopher Thomas Hobbes, in *Leviathan*: "Nature (the Art whereby God hath made and governes the World) is by the *Art* of man, as in many other things, so in this also imitated, that it can make an Artificial Animal. For seeing life is but a motion of Limbs, the begining whereof is in some principall part within; why may we not say, that all *Automata* (Engines that move themselves by springs and wheeles as doth a watch) have an artificiall life?" (1651:81).

6. For an ethnography of an evolutionary robotics center in Europe, see Risan 1996.

7. I am sweeping a few things under the continent. There are a series of conferences on the simulation of adaptive behavior hosted by yet another transnational set of folks, many of whom are Europe-based and who have primary interests in robotics. These conferences have been held in France, the United Kingdom, and the United States.

8. George Kampis, shortly before the workshop, sent out an email on autopoiesis. He concluded that there can be models of autopoiesis but that these cannot themselves be considered autopoietic: "There can be no such thing as an algorithm for an autopoietic system. Algorithms are prototyped by machines where there is no feedback from the outputs towards the defining primitives. No formal system can depend on axioms that it produces as theorems. An algorithm is, by definition, an 'allopoietic' system, one that produces something else than itself, whereas an 'autopoietic' system should be the other way around." This contention that models of living things cannot become living things themselves is in direct opposition to the computationalist program of Langton and others.

9. Doyle argues that the incoherence and multiplicity of "life" is a powerful resource for imagining artificial life at all. The infusion of "life" into various pockets of the world is reminiscent of what Keller has described as the leaking of the "gene" into an increasingly cybernetic organism. Keller argues that the coherence of the "gene" is falling apart with the discovery of such phenomena as split, overlapping, and nested genes and with growing consciousness that the "coding" functions of the gene are dispersed from DNA into complex relations

between DNA, RNA, and often, regulatory processes operating on the level of protein synthesis which can change DNA transcripts themselves (Keller 1996). The obsession with the "gene" that has animated molecular biology has paradoxically drawn attention back to the organisms in which genes operate. Artificial Life obsession with "life" has similarly made vitality disappear, leaving it to saturate everything, forcing a recognition that life only exists in material relations of interpretation.

10. My conference paper was a permutation of an SFI technical report I had written a few months previously, a report that enjoyed a fair amount of controversy. After I had been in residence at the Institute for a while, a few scientists encouraged me to make use of the in-house mechanism for publishing preliminary research results. I patched together a short article, emphasizing that this was work in progress and that I would be happy to discuss my approach and conclusions. After the paper had been on the shelves for a week, after about 80 of the 100 copies had made it into circulation, and after it had prompted many stimulating discussions, it was removed by the administration, who found its contents objectionable. I was told that my arguments were not subject to empirical verification, were "not scientific." I was also told that the paper was pulled because the administration did not want members of the science board, the board of trustees, or the Santa Fe Institute business network approaching them and asking, "Is this what my money is funding?" Later, I was told that the paper would not be considered for republication until it was "approved." The administrators who made this decision—people trained, for the most part, in physics—said they felt torn between the need to protect academic freedom and the need to make sure that things were held to "scientific standards." I was told by a secretary that the undistributed copies of the paper had been "recycled." This framing of the matter as a question of disposing of paper in an ecologically correct way skirted what many people saw as the issue: censorship. Many scientists and staff thought the decision to remove my paper was a poor one and did not share the administration's objections. The censorship was justified by an appeal to scientific standards, standards that were nowhere defined for me to use as a guideline, standards that, as far as I could see, were not being enforced for other papers, particularly those that indulged in speculations about ultimate human nature or economic rationality. The Artificial Life community intersects with but is by no means identical to the Institute, as is evidenced by the fact that I was invited to give a version of the paper at the Artificial Life conference I discuss here.

11. "Life" occupied a central place in Pope John Paul II's March 1995 encyclical letter, "Evangelium Vitae" (Gospel of Life), in which he declared that scientific and technological advancements were enabling the perpetration of "crimes against life" (by which he meant abortion and euthanasia). Notably, the pope referred only fleetingly to "nature" or "natural law," choosing to base his argument on morality. Given that he accepted the efforts of medicine to protect "life," the pope felt comfortable mixing the categories of culture and nature and did not find it useful to ground moral claims in "nature." This must be seen as

quite different from the way earlier Christian ideas about morality sought above all a "natural" warrant.

12. Benjamin argues that films achieve effects of naturalness only through the greatest artifice, the artifice that erases its own presence: "The equipment-free aspect of reality here has become the height of artifice; the sight of immediate reality has become an orchid in the land of technology" (1936:233).

13. There is a growing literature on drag in many of its multiple forms, and I cannot hope to survey it here. A good starting point for cultural studies approaches to the topic is Garber's *Vested Interests* (1992). Garber deals with the different racial, historical, and national logics and power dynamics of many kinds of drag, including gay and straight male-to-female drag, straight and lesbian female-to-male drag (as well as variations and contestations of the butch-femme theme), and drag that shades into transsexuality. In newer literature, the boundaries between transvestism and transsexuality have become a focus of analysis, and in some places the word *transgender* has been taken up to speak to how "anatomical sex" and gendered ways of dressing and acting have entered into new combinations (see MacKenzie 1994). In some treatments of the transformation of bodies through medical intervention and wardrobe modification, what is "natural" is becoming a less important question, especially as some people have chosen to remain sex/gender ambiguous or to occupy a body that is a pastiche of differently "sexed" parts (see Smyth 1995).

14. We could follow Franklin, who, in her discussion of how the new genetics undermines stable genderings of "nature" and "life," writes that "reproduction in the context of the new genetic sciences is not so much 'enterprised-up' as it is 'camped-up,' in a mocking re-presentation of 'the real thing' it simulates" (Franklin 1995:71–72). ("Enterprised-up" is Strathern's phrase characterizing what happens to "nature" and "reproduction" with the advent of technologies that routinely mix nature and culture.)

15. Though one might say that the definition of cognition has been changing in order to outmaneuver the claims of those who argue that machines can think (see Dreyfus 1979; Turkle 1991). For contrary arguments—that computers are now generally considered intelligent—see Turkle 1995; Boden 1996.

References

Ackley, David. 1997. "ccr: A Network of Worlds for Research." In *Artificial Life V*, ed. Christopher G. Langton and Katsunori Shimohara, 116–123. Cambridge, Mass.: MIT Press/Bradford Books.

Ackley, David, and Michael Littman. 1992a. "Interactions between Learning and Evolution." In *Artificial Life II*, ed. Christopher G. Langton, Charles Taylor, J. Doyne Farmer, and Steen Rasmussen, 487–509. Redwood City, Calif.: Addison-Wesley.

———. 1992b. "Learning from Natural Selection in an Artificial Environment." *Artificial Life II Video Proceedings*, ed. Christopher G. Langton. Redwood City, Calif.: Addison-Wesley.

Anderson, Philip, Kenneth Arrow, and David Pines, eds. 1988. *The Economy as an Evolving Complex System*. Redwood City, Calif.: Addison-Wesley.

Ansell Pearson, Keith. 1997. *Viroid Life: Perspectives on Nietzsche and the Transhuman Condition*. London: Routledge.

Aristotle. 1979. *Generation of Animals*. Translated from the ancient Greek by A. L. Peck. Loeb Classical Library. Cambridge, Mass.: Harvard University Press.

Asad, Talal. 1993. *Genealogies of Religion: Discipline and Reasons of Power in Christianity and Islam*. Baltimore: Johns Hopkins University Press.

Baldwin, James. 1984. "On Being 'White' . . . And Other Lies." *Essence*, April, 90, 92.

Barinaga, Marcia. 1994. "Surprises Across the Cultural Divide." *Science* 263:1468–1472.

Baudrillard, Jean. 1983. *Simulations*. Translated from the French by Paul Foss, Paul Patton, and Philip Beitchman. New York: Semiotext(e).

Bedau, Mark. 1996. "The Nature of Life." In *The Philosophy of Artificial Life*, ed. Margaret A. Boden, 332–357. New York: Oxford University Press.

Bedau, Mark, and Norman Packard. 1992. "Measurement of Evolutionary Activity, Teleology, and Life." In *Artificial Life II*, ed. Christopher G. Langton,

Charles Taylor, J. Doyne Farmer, and Steen Rasmussen, 431–461. Redwood City, Calif.: Addison-Wesley.

Belenky, Mary Field, Blythe McVicker Clinchy, Nancy Rule Goldberger, and Jill Mattuck Tarule. 1986. *Women's Ways of Knowing: The Development of Self, Voice, and Mind.* New York: Basic Books.

Bellah, Robert N. 1992. *The Broken Covenant: American Civil Religion in Time of Trial,* 2d ed. Chicago: University of Chicago Press. First edition 1975.

Benjamin, Walter. 1936. "The Work of Art in the Age of Mechanical Reproduction." Translated from the German by Harry Zohn. In *Illuminations,* ed. Hannah Arendt, 217–251. New York: Schocken Books, 1968.

The Bible. King James Version. n.p.: World Bible Publishers, Inc.

The New Jerusalem Bible. New York: Doubleday, 1985.

Bleecker, Julian. 1995. "Urban Crisis: Past, Present, and Virtual." *Socialist Review* 24(1–2):189–221.

Bleier, Ruth. 1984. *Science and Gender: A Critique of Biology and Its Theories on Women.* New York: Pergamon Press.

Boden, Margaret A., ed. 1996. *The Philosophy of Artificial Life.* New York: Oxford University Press.

Booker, Lashon B. 1985. "Improving the Performance of Genetic Algorithms in Classifier Systems." In *Proceedings of the First International Conference on Genetic Algorithms,* ed. John J. Grefenstette, 80–92. Hillsdale, N.J.: Lawrence Erlbaum.

Boston Women's Health Book Collective. 1992. *The New Our Bodies, Ourselves.* New York: Simon and Schuster. First edition 1969.

Bourdieu, Pierre. 1972. *Outline of a Theory of Practice.* Translated from the French by Richard Nice. Cambridge: Cambridge University Press, 1977.

———. 1975. "The Specificity of the Scientific Field and the Social Conditions of the Progress of Reason." *Social Science Information* 14(6):19–47.

Bremer, Michael. 1992. *SimLife User Manual.* Orinda, Calif.: Maxis.

Brockman, John, ed. 1995. *The Third Culture: Beyond the Scientific Revolution.* New York: Simon and Schuster.

Brown, Michael F. 1997. *The Channeling Zone: American Spirituality in an Anxious Age.* Cambridge, Mass.: Harvard University Press.

Burian, Richard M., and Robert C. Richardson. 1996. "Form and Order in Evolutionary Biology." In *The Philosophy of Artificial Life,* ed. Margaret A. Boden, 146–172. New York: Oxford University Press.

Burke, Gibbons. 1993. "Good Trading a Matter of Breeding?" *Futures* 12(5):26–29.

Butler, Judith. 1989. *Gender Trouble: Feminism and the Subversion of Identity.* New York: Routledge.

———. 1993. *Bodies that Matter: On the Discursive Limits of "Sex."* New York: Routledge.

Butler, Samuel. 1872. *Erewhon, or Over the Range.* London: Trübner.

Bynum, Caroline Walker. 1989. "The Female Body and Religious Practice in the Later Middle Ages." In *Fragments for a History of the Human Body, Part One,* ed. Michel Feher with Ramona Naddaff and Nadia Tazi, 160–219. New York: Zone.

Capra, Fritjof. 1975. *The Tao of Physics: An Exploration of the Parallels between Modern Physics and Eastern Mysticism*. Boulder, Colo.: Shambala.

"Career Guide for Women in Science." Compiled by a group of women faculty and professionals at Cornell, for the occasion of a Conference for Women in Science. February 20, 1988.

Carroll, Lewis. 1871. "Through the Looking-Glass." In *Alice's Adventures in Wonderland and Through the Looking-Glass*. New York: Bantam, 1988.

Casper, Monica. 1995. "Fetal Cyborgs and Technomoms on the Reproductive Frontier: Which Way to the Carnival?" In *The Cyborg Handbook,* ed. Chris Hables Gray with the assistance of Heidi J. Figueroa-Sarriera and Steven Mentor, 183–202. New York: Routledge.

Casti, John. 1997. *Would-be Worlds: How Simulation Is Changing the Frontiers of Science*. New York: Wiley.

Channell, David. 1991. *The Vital Machine: A Study of Technology and Organic Life*. New York: Oxford University Press.

Churchill, Ward. 1994. *Indians Are Us? Culture and Genocide in Native North America*. Monroe, Maine: Common Courage Press.

Clarke, Adele. 1995. "Modernity, Postmodernity, & Reproductive Processes, ca. 1890–1990, or 'Mommy, Where Do Cyborgs Come from Anyway?' " In *The Cyborg Handbook,* ed. Chris Hables Gray, with the assistance of Heidi J. Figueroa-Sarriera and Steven Mentor, 139–155. New York: Routledge.

Cleveland, Gary A., and Stephen F. Smith. 1989. "Using Genetic Algorithms to Schedule Shop Flow Releases." In *Proceedings of the Third International Conference on Genetic Algorithms,* ed. J. David Schaffer, 160–179. San Mateo, Calif.: Morgan Kaufmann.

Clynes, Manfred E., and Nathan S. Kline. 1960. "Cyborgs and Space." *Astronautics*, September, 26–27, 75–76.

Cohen, John. 1966. *Human Robots in Myth and Science*. New York: A. S. Barnes.

Cohn, Carol. 1987. "Sex and Death in the Rational World of Defense Intellectuals." *Signs* 12(4):687–718.

Collier, Jane, and Sylvia Yanagisako. 1987. "Toward a Unified Analysis of Gender and Kinship." In *Toward a Unified Analysis of Gender and Kinship,* ed. Jane Collier and Sylvia Yanagisako, 14–50. Stanford: Stanford University Press.

Collier, Jane, Bill Maurer, and Liliana Suárez Navaz. 1995. "Sanctioned Identities: Legal Constructions of Modern Personhood." *Identities* 2(1–2):1–27.

Collins, Robert, and David Jefferson. 1992. "AntFarm: Towards Simulated Evolution." In *Artificial Life II,* ed. Christopher G. Langton, Charles Taylor, J. Doyne Farmer, and Steen Rasmussen, 579–602. Redwood City, Calif.: Addison-Wesley.

Comaroff, Jean, and John Comaroff. 1991. *Of Revelation and Revolution: Christianity, Colonialism, and Consciousness in South Africa*. Chicago: University of Chicago Press.

Corcoran, Elizabeth. 1992. "The Edge of Chaos: Complexity Is a Metaphor at the Santa Fe Institute." *Scientific American* 267(4):17–22.

Cowan, George. 1990. "President's Message." *Bulletin of the Santa Fe Institute* 5(2):1.

Crowe, Michael J. 1986. *The Extraterrestrial Life Debate, 1750–1900: The Idea of a Plurality of Worlds from Kant to Lowell.* Cambridge: Cambridge University Press.

Crutchfield, James, and Melanie Mitchell. 1994. "The Evolution of Emergent Computation." SFI preprint 94-03-012.

Davis, Lawrence, ed. 1996. *Handbook of Genetic Algorithms.* London: International Thomson Computer Press.

Dawkins, Richard. 1976. *The Selfish Gene.* Oxford: Oxford University Press.

———. 1986. *The Blind Watchmaker.* New York: W. W. Norton.

———. 1989. "The Evolution of Evolvability." In *Artificial Life,* ed. Christopher G. Langton, 201–220. Redwood City, Calif.: Addison-Wesley.

———. 1992. "Progress." In *Keywords in Evolutionary Biology,* ed. Evelyn Fox Keller and Elisabeth A. Lloyd, 263–272. Cambridge, Mass.: Harvard University Press.

DeBouzek, Jeannette, and Diane Reyna. 1992. *Gathering Up Again: Fiesta in Santa Fe.* Quotidian Independent Documentary Research. Videocassette.

DeLanda, Manuel. 1992. "Nonorganic Life." In *Incorporations,* ed. Jonathan Crary and Sanford Kwinter, 128–167. New York: Zone.

Delaney, Carol. 1986. "The Meaning of Paternity and the Virgin Birth Debate." *Man* 21(3):494–513.

———. 1991. *The Seed and the Soil: Gender and Cosmology in Turkish Village Society.* Berkeley: University of California Press.

de Lauretis, Teresa. 1987. *Technologies of Gender: Essays on Theory, Film, and Fiction.* Bloomington: Indiana University Press.

Dick, Steven J. 1982. *Plurality of Worlds: The Origins of the Extraterrestrial Life Debate from Democritus to Kant.* Cambridge: Cambridge University Press.

Douglas, Mary. 1966. *Purity and Danger: An Analysis of the Concepts of Pollution and Taboo.* London: Routledge & Kegan Paul.

Downey, Gary Lee, and Joseph Dumit, eds. 1998. *Cyborgs and Citadels: Anthropological Interventions in Emerging Sciences and Technologies.* Santa Fe, New Mex.: School of American Research Press.

Downey, Gary Lee, Joseph Dumit, and Sarah Williams. 1995. "Cyborg Anthropology." *Cultural Anthropology* 10(2):264–269.

Doyle, Richard. 1992. "The Rhetorical Software of Artificial Life." Paper presented at Artificial Life III, Santa Fe, New Mex., June 15–19.

———. 1993. "It's All in the Genes: At Play in the Phenotype of Artificial Life." Manuscript.

———. 1996. " 'More Life, Fucker': Or, the Anticipatory Effects of Simulacra." Paper presented at Simulating Knowledge: Cultural Analysis of Computer Modeling in the Life Sciences, Cornell University, April 19–21.

———. 1997a. "Artificial Life Support: Nodes in the ALife Ribotype." Paper presented at Growing Explanations: Historical Perspectives on Recent Scientific Practice, Princeton University, February 15–16.

———. 1997b. *On Beyond Living: Rhetorical Transformations of the Life Sciences.* Stanford: Stanford University Press.

Dreyfus, Hubert L. 1979. *What Computers Can't Do: A Critique of Artificial Reason,* 2d ed. New York: Harper and Row. First edition 1972.

Duden, Barbara. 1993. "Visualizing 'Life.'" *Science as Culture* 3(4):562–600.

Dupré, John. 1990. "Global versus Local Perspectives on Sexual Difference." In *Theoretical Perspectives on Sexual Difference,* ed. Deborah Rhode, 33–46. Stanford: Stanford University Press.

Easlea, Brian. 1983. *Fathering the Unthinkable: Masculinity, Scientists, and the Arms Race.* London: Pluto Press.

Easton, Fred. F., and Nashat Mansour. 1993. "A Distributed Genetic Algorithm for Employee Staffing and Scheduling Problems." In *Proceedings of the Fifth International Conference on Genetic Algorithms,* ed. Stephanie Forrest, 360–367. San Mateo, Calif.: Morgan Kaufmann.

Edwards, Paul. 1996. *The Closed World: Computers and the Politics of Discourse in Cold War America.* Cambridge, Mass.: MIT Press.

Eglash, Ronald. 1995. "African Influences in Cybernetics." In *The Cyborg Handbook,* ed. Chris Hables Gray, with the assistance of Heidi J. Figueroa-Sarriera and Steven Mentor, 17–27. New York: Routledge.

———. Forthcoming. *African Fractals: Traditional Culture and Modern Computing.* New Brunswick: Rutgers University Press.

Emmeche, Claus. 1991. *The Garden in the Machine: The Emerging Science of Artificial Life.* Translated from the Danish by Steven Sampson. Princeton: Princeton University Press, 1994.

———. 1992. "Life as an Abstract Phenomenon: Is Artificial Life Possible?" In *Toward a Practice of Autonomous Systems: Proceedings of the First European Conference on Artificial Life,* ed. Francisco Varela and Paul Bourgine, 466–474. Cambridge, Mass.: MIT Press/Bradford Books.

———. 1994. "Life as a Multiverse Phenomenon: the Biosemiotics of Computation." In *Artificial Life III,* ed. Christopher G. Langton, 553–568. Redwood City, Calif.: Addison-Wesley.

———. 1997. "Constructing and Explaining Emergence in Artificial Life: Can Life Be Defined as an Emergent Phenomenon?" Paper presented at Growing Explanations: Historical Perspectives on Recent Scientific Practice, Princeton University, February 15–16.

Epstein, Joshua, and Robert Axtell. 1996. *Growing Artificial Societies: Social Science from the Bottom Up.* Cambridge, Mass.: MIT Press.

Escobar, Arturo. 1995. *Encountering Development: The Making and Unmaking of the Third World.* Princeton: Princeton University Press.

Eshelman, Larry J, and J. David Schaffer. 1991. "Preventing Premature Convergence in Genetic Algorithms by Preventing Incest." In *Proceedings of the Fourth International Conference on Genetic Algorithms,* ed. Richard K. Belew and Lashon B. Booker, 115–122. San Mateo, Calif.: Morgan Kaufmann.

———. 1993. "Crossover's Niche." In *Proceedings of the Fifth International Conference on Genetic Algorithms,* ed. Stephanie Forrest, 9–14. San Mateo, Calif.: Morgan Kaufmann.

Farmer, J. Doyne, and Alletta d'A. Belin. 1992. "Artificial Life: The Coming Evolution." In *Artificial Life II,* ed. Christopher G. Langton, Charles Taylor, J. Doyne Farmer, and Steen Rasmussen, 815–840. Redwood City, Calif.: Addison-Wesley.

Fausto-Sterling, Anne. 1992. *Myths of Gender,* rev. ed. New York: Basic Books. First edition 1985.

Fernández-Kelly, María Patricia. 1983. *For We Are Sold, I and My People: Women and Industry in Mexico's Frontier.* Albany: State University of New York Press.

Fernández Ostolaza, Julio, and Álvaro Moreno Bergareche. 1992. *Vida artificial.* Madrid: Eudema.

———. 1997. *La vie artificielle.* Translated from the Spanish by Mylène de Fabrique Saint-Tours and Patricia Rey. Paris: Éditions du Seuil.

Flower, Michael, and Deborah Heath. 1993. "Micro-Anatomo Politics: Mapping the Human Genome Project." In *Bio-Politics: The Anthropology of the New Genetics and Immunology,* ed. Deborah Heath and Paul Rabinow. Special issue, *Culture, Medicine, and Psychiatry* 17:27–41.

Fontana, Walter, and Leo Buss. 1994. "What Would Be Conserved if 'The Tape Were Played Twice'?" In *Complexity: Metaphors, Models, and Reality,* ed. George Cowan, David Pines, and David Meltzer, 223–244. Redwood City, Calif.: Addison-Wesley.

Forrest, Stephanie. 1993. "Genetic Algorithms: Principles of Natural Selection Applied to Computation." *Science* 261:872–878.

Foucault, Michel. 1966. *The Order of Things: An Archaeology of the Human Sciences.* Translated from the French *Les mots et los choses.* New York: Random House, 1970.

———. 1976. *The History of Sexuality,* vol. 1. Translated from the French by Robert Hurley. New York: Vintage, 1978.

Fox, Robin. 1967. *Kinship and Marriage: An Anthropological Perspective.* Harmondsworth: Penguin.

Frankenberg, Ruth. 1993. *White Women, Race Matters: The Social Construction of Whiteness.* Minneapolis: University of Minnesota Press.

Franklin, Sarah. 1993a. "Life Itself." Paper prepared for the Detraditionalisation conference, Centre for Cultural Values, Lancaster University, Lancaster, England, June 9.

———. 1993b. "Postmodern Procreation: Representing Reproductive Practice." *Science as Culture* 3(4):522–561.

———. 1995. "Romancing the Helix: Nature and Scientific Discovery." In *Romance Revisited,* ed. L. Pearce and J. Stacey, 63–77. London: Lawrence and Wishart.

———. 1997. *Embodied Progress: A Cultural Account of Assisted Conception.* London: Routledge.

Fraser, Nancy, and Linda Gordon. 1992. "Contract versus Charity: Why Is There No Social Citizenship in the United States?" *Socialist Review* 22(3):45–67.

Fujimura, Joan. 1996. *Crafting Science: A Sociology of the Quest for the Genetics of Cancer.* Cambridge, Mass.: Harvard University Press.

Galison, Peter. 1996. "Computer Simulations and the Trading Zone." In *The Disunity of Science: Boundaries, Contexts, and Power,* ed. Peter Galison and David J. Stump, 118–157. Stanford: Stanford University Press.

———. 1997. *Image and Logic: A Material Culture of Microphysics.* Chicago: University of Chicago Press.

Garber, Marjorie. 1992. *Vested Interests: Cross-Dressing and Cultural Anxiety.* New York: Routledge.

Geertz, Clifford. 1973. "Religion as a Cultural System." In *The Interpretation of Cultures,* 87–125. New York: Basic Books.

Geertz, Hildred, and Clifford Geertz. 1975. *Kinship in Bali.* Chicago: University of Chicago Press.

Gell-Mann, Murray. 1987. "The Concept of the Institute." In *Emerging Syntheses in Science,* ed. David Pines, 1–15. Redwood City, Calif.: Addison-Wesley.

———. 1994. *The Quark and the Jaguar: Adventures in the Simple and the Complex.* New York: W. H. Freeman.

Genova, Judith. 1989. "Women and the Mismeasure of Thought." In *Feminism and Science,* ed. Nancy Tuana, 211–227. Bloomington: Indiana University Press.

Gibson, William. 1984. *Neuromancer.* New York: Ace.

Gilligan, Carol. 1982. *In a Different Voice: Psychological Theory and Women's Development.* Cambridge, Mass.: Harvard University Press.

Gilroy, Paul. 1993. *The Black Atlantic: Modernity and Double Consciousness.* Cambridge, Mass.: Harvard University Press.

Gleick, James. 1987. *Chaos: Making a New Science.* New York: Penguin.

Glosserman, Brad. 1994. "The Scientific Mavericks of Santa Fe Zero in on Evolution." *Japan Times,* March 16, 15.

Goethe, Johann Wolfgang. 1832. *Faust.* Translated from the German by Philip Wayne. Harmondsworth: Penguin, 1949.

Goldberg, David E. 1989a. *Genetic Algorithms in Search, Optimization, and Machine Learning.* Reading, Mass.: Addison-Wesley.

———. 1989b. "Zen and the Art of Genetic Algorithms." In *Proceedings of the Third International Conference on Genetic Algorithms,* ed. J. David Schaffer, 80–85. San Mateo, Calif.: Morgan Kaufmann.

Goodwin, Brian. 1994a. "Developmental Complexity and Evolutionary Order." In *Complexity: Metaphors, Models, and Reality,* ed. George Cowan, David Pines, and David Meltzer, 205–222. Redwood City, Calif.: Addison-Wesley.

———. 1994b. *How the Leopard Changed Its Spots: The Evolution of Complexity.* New York: Charles Scribner's Sons.

Gould, Stephen Jay. 1979. *Ever Since Darwin: Reflections in Natural History.* New York: W. W. Norton.

———. 1989. *Wonderful Life: The Burgess Shale and the Nature of History.* New York: W. W. Norton.

———. 1995. " 'What is Life?' as a Problem in History." In *What Is Life? The Next Fifty Years: Speculations on the Future of Biology,* ed. Michael Murphy and Luke O'Neill, 25–39. Cambridge: Cambridge University Press.

Gramsci, Antonio. 1971. *Selections from the Prison Notebooks of Antonio Gramsci.* Edited and translated from the Italian by Quintin Hoare and Geoffrey Nowell Smith. London: Lawrence and Wishart, originally written 1929–1937.

Gray, Chris Hables, ed., with the assistance of Heidi J. Figueroa-Sarriera and Steven Mentor. 1995. *The Cyborg Handbook.* New York: Routledge.

Gruau, Frédéric. 1993. "Genetic Synthesis of Modular Neural Networks." In *Proceedings of the Fifth International Conference on Genetic Algorithms,* ed. Stephanie Forrest, 312–317. San Mateo, Calif.: Morgan Kaufmann.

Gupta, Akhil, and James Ferguson. 1992. "Beyond 'Culture': Space, Identity, and the Politics of Difference." *Cultural Anthropology* 7(1):6–23.

Gusterson, Hugh. 1996. *Nuclear Rites: A Weapons Laboratory at the End of the Cold War.* Berkeley: University of California Press.

Halberstam, Judith. 1991. "Automating Gender: Postmodern Feminism in the Age of the Intelligent Machine." *Feminist Studies* 17(3):439–460.

Hamilton, W. D. 1964. "The Genetical Evolution of Social Behavior." *Journal of Theoretical Biology* 7:1–51.

Haney López, Ian F. 1996. *White by Law: The Legal Construction of Race.* New York: New York University Press.

Haraway, Donna. 1989. *Primate Visions: Gender, Race, and Nature in the World of Modern Science.* New York: Routledge.

———. 1991a. "Animal Sociology and a Natural Economy of the Body Politic: A Political Physiology of Dominance." In *Simians, Cyborgs, and Women: The Reinvention of Nature,* 7–20. New York: Routledge.

———. 1991b. "The Biological Enterprise: Sex, Mind, and Profit from Human Engineering to Sociobiology." In *Simians, Cyborgs, and Women: The Reinvention of Nature,* 43–68. New York: Routledge.

———. 1991c. "The Biopolitics of Postmodern Bodies: Constitutions of Self in Immune System Discourse." In *Simians, Cyborgs, and Women: The Reinvention of Nature,* 203–230. New York: Routledge.

———. 1991d. "A Cyborg Manifesto: Science, Technology, and Socialist-Feminism in the Late Twentieth Century." In *Simians, Cyborgs, and Women: The Reinvention of Nature,* 149–182. New York: Routledge.

———. 1991e. "The Promises of Monsters: A Regenerative Politics for Inappropriate/d Others." In *Cultural Studies,* ed. Lawrence Grossberg, Gary Nelson, and Paula Treichler, 295–337. New York: Routledge.

———. 1991f. "Situated Knowledges: The Science Question in Feminism and the Privilege of Partial Perspective." In *Simians, Cyborgs, and Women: The Reinvention of Nature,* 183–202. New York: Routledge.

———. 1994. "A Game of Cat's Cradle: Science Studies, Feminist Theory, Cultural Studies." *Configurations* 1:59–71.

———. 1997. *Modest_Witness@Second_Millennium.FemaleMan©_Meets_OncoMouse™: Feminism and Technoscience.* New York: Routledge.

Harding, Sandra. 1986. *The Science Question in Feminism.* Ithaca: Cornell University Press.

———. 1991. *Whose Science? Whose Knowledge? Thinking from Women's Lives.* Ithaca: Cornell University Press.

Harding, Sandra, ed. 1993. *The "Racial" Economy of Science: Toward a Democratic Future.* Bloomington: Indiana University Press.

Harnad, Stevan. 1994. "Levels of Functional Equivalence in Reverse Bioengineering." *Artificial Life* 1(3):293–301.

Harp, Steven Alex, Tariq Samad, and Aloke Guha. 1989. "Toward the Genetic Synthesis of Neural Networks." In *Proceedings of the Third International Conference on Genetic Algorithms,* ed. J. David Schaffer, 360–369. San Mateo, Calif.: Morgan Kaufmann.

Harris, Olivia, and Kate Young. 1981. "Engendered Structures: Some Problems in the Analysis of Reproduction." In *The Anthropology of Pre-Capitalist Societies,* ed. Joel S. Kahn and Josep R. Llobera, 109–147. London: Macmillan.

Hartouni, Valerie. 1997. *Cultural Conceptions: On Reproductive Technologies and the Remaking of Life.* Minneapolis: University of Minnesota Press.

Harvey, David. 1989. *The Condition of Postmodernity: An Enquiry into the Origins of Cultural Change.* Cambridge, Mass.: Blackwell.

Hayden, Corinne. 1998. "A Biodiversity Sampler for the Millennium." In *Reproducing Reproduction: Kinship, Power, and Technological Innovation,* ed. Sarah Franklin and Helena Ragoné, 173–206. Philadelphia: University of Pennsylvania Press.

Hayles, N. Katherine. 1994a. "The Closure of Artificial Worlds: How Nature Became Virtual." Paper presented at Vital Signs: Cultural Perspectives on Coding Life and Vitalizing Code, Stanford University, June 2–4.

———. 1994b. "Narratives of Evolution and the Evolution of Narratives." In *Cooperation and Conflict in General Evolutionary Processes,* ed. John L. Casti and Anders Karlqvist, 113–132. New York: Wiley.

———. 1995. "Simulated Nature and Natural Simulations: Rethinking the Relation between the Beholder and the World." In *Uncommon Ground: Toward the Reinvention of Nature,* ed. William Cronon, 409–425. New York: W. W. Norton.

———. 1996. "Narratives of Artificial Life." In *FutureNatural: Nature, Science, Culture,* ed. George Robertson, Melinda Mash, Lisa Tucker, Jon Bird, Barry Curtis, and Tim Putnam, 146–164. London: Routledge.

Hazen-Hammond, Susan. 1988. *A Short History of Santa Fe.* San Francisco: Lexikos.

Healy, Dave. 1997. "Cyberspace and Place: The Internet as Middle Landscape on the Electronic Frontier." In *Internet Culture,* ed. David Porter, 55–68. New York: Routledge.

Helmreich, Stefan. 1992. "The Historical and Epistemological Ground of von Neumann's Theory of Self-Reproducing Automata and Theory of Games." In *Toward a Practice of Autonomous Systems: Proceedings of the First European Conference on Artificial Life,* ed. Francisco Varela and Paul Bourgine, 385–391. Cambridge, Mass.: MIT Press/Bradford Books.

Herrigel, Eugen. 1948. *Zen in the Art of Archery.* Translated from the German by R. F. C. Hull. New York: Pantheon, 1953.

Hess, David. 1992. "Introduction: The New Ethnography and the Anthropology of Science and Technology." In *Knowledge and Society 9: The Anthropology of Science and Technology*, ed. David Hess and Linda Layne, 1–26. Greenwich, Conn.: JAI Press.

———. 1993. *Science in the New Age: The Paranormal, Its Defenders and Debunkers, and American Culture.* Madison: University of Wisconsin Press.

Hiebeler, David. 1993. "Implications of Creation." SFI preprint 93-05-025.

Hobbes, Thomas. 1651. *Leviathan.* Harmondsworth: Penguin, 1985.

Hofstadter, Douglas R. 1979. *Gödel, Escher, Bach: An Eternal Golden Braid.* New York: Vintage.

Hogeweg, Paulien. 1989. "Mirror Beyond Mirror: Puddles of Life." In *Artificial Life,* ed. Christopher G. Langton, 297–316. Redwood City, Calif.: Addison-Wesley.

Holland, John. 1975. *Adaptation in Natural and Artificial Systems.* Ann Arbor: University of Michigan Press.

———. 1988. "The Global Economy as an Adaptive Process." In *The Economy as an Evolving Complex System,* ed. Philip Anderson, Kenneth Arrow, and David Pines, 117–124. Redwood City, Calif.: Addison-Wesley.

———. 1992a. *Adaptation in Natural and Artificial Systems,* 2d ed. Cambridge, Mass.: MIT Press/Bradford Books.

———. 1992b. "Genetic Algorithms." *Scientific American* 267(1):66–72.

———. 1993. "The ECHO Model." In *1992 Annual Report on Scientific Programs.* Santa Fe, New Mex.: Santa Fe Institute.

———. 1994. "Echoing Emergence: Objectives, Rough Definitions, and Speculations for ECHO-Class Models." In *Complexity: Metaphors, Models, and Reality,* ed. George Cowan, David Pines, and David Meltzer, 309–342. Redwood City, Calif.: Addison-Wesley.

———. 1995. *Hidden Order: How Adaptation Builds Complexity.* Reading, Mass.: Addison-Wesley.

Holldobler, Bert, and Edward O. Wilson. 1990. *The Ants.* Cambridge, Mass.: Harvard University Press.

Holloway, Marguerite. 1993. "A Lab of Her Own." *Scientific American* 269(5):94–103.

hooks, bell. 1992a. "Representations of Whiteness in the Black Imagination." In *Black Looks: Race and Representation,* 165–178. Boston: South End Press.

———. 1992b. "Selling Hot Pussy: Representations of Black Female Sexuality in the Cultural Marketplace." In *Black Looks: Race and Representation,* 61–78. Boston: South End Press.

Irigaray, Luce. 1977. "This Sex Which Is Not One." Translated from the French by Claudia Reader. In *New French Feminisms,* ed. Elaine Marks and Isabelle de Courtivron, 99–110. New York: Shocken Books, 1980.

Ito, Mizuko. 1997. "Virtuality Embodied: The Reality of Fantasy in a Multi-User Dungeon." In *Internet Culture,* ed. David Porter, 87–109. New York: Routledge.

Jaggar, Alison M. 1983. *Feminist Politics and Human Nature*. Brighton, U.K.: Harvester Press.

Jaimes, M. Annette. 1992. "Federal Indian Identification Policy: A Usurpation of Indigenous Sovereignty in North America." In *The State of Native America: Genocide, Colonization, and Resistance*, ed. M. Annette Jaimes, 123–138. Boston: South End Press.

Jameson, Frederic. 1991. *Postmodernism, or, The Cultural Logic of Late Capitalism*. Durham: Duke University Press.

Jefferson, David, Robert Collins, Claus Cooper, Michael Dyer, Margot Flowers, Richard Korf, Charles Taylor, and Alan Wang. 1992. "Evolution as a Theme in Artificial Life: The Genesys/Tracker System." In *Artificial Life II*, ed. Christopher G. Langton, Charles Taylor, J. Doyne Farmer, and Steen Rasmussen, 549–578. Redwood City, Calif.: Addison-Wesley.

Jenks, Andrew. 1994. "Datadance with a Double Helix: Studying Complex Problems with Genetic Math." Unpaginated reprint from *Washington Technology* 9(2).

Pope John Paul II. 1995. "Encyclical Letter: Evangelium Vitae, Addressed by the Supreme Pontiff John Paul II to the Bishops, Priests, and Deacons, Men and Women Religious, Lay Faithful, and All People of Good Will on the Value and Inviolability of Human Life."

Jones, Terry, and Stephanie Forrest. 1993. "An Introduction to SFI Echo." SFI preprint 93-12-074.

Kampis, George. 1995. "The Inside and Outside Views of Life." In *Advances in Artificial Life: The Third European Conference on Artificial Life* (Lecture Notes in Artificial Intelligence 929, Subseries of Lecture Notes in Computer Science), ed. F. Morán, A. Moreno, J. J. Morelo, and P. Chacón, 95–102. Berlin: Springer.

Kandinsky, Wassily. 1911. *Concerning the Spiritual in Art*. Translated from the German by M. T. H. Sadler. New York: Dover Publications, 1977.

Kauffman, Stuart A. 1993. *Origins of Order: Self-Organization and Selection in Evolution*. New York: Oxford University Press.

———. 1995. " 'What Is Life?' Was Schrödinger Right?" In *What Is Life? The Next Fifty Years: Speculations on the Future of Biology*, ed. Michael Murphy and Luke O'Neill, 83–114. Cambridge: Cambridge University Press.

Kaye, Howard L. 1986. *The Social Meaning of Modern Biology: From Social Darwinism to Sociobiology*. New Haven: Yale University Press.

Keeley, Brian. 1994. "Against the Global Replacement: On the Application of the Philosophy of Artificial Intelligence to Artificial Life." In *Artificial Life III*, ed. Christopher G. Langton, 569–587. Redwood City, Calif.: Addison-Wesley.

Keller, Evelyn Fox. 1983. *A Feeling for the Organism: The Life and Work of Barbara McClintock*. New York: W. H. Freeman.

———. 1985. *Reflections on Gender and Science*. New Haven: Yale University Press.

———. 1992a. "Demarcating Public from Private Values in Evolutionary Discourse." In *Secrets of Life, Secrets of Death: Essays on Language, Gender and Science*, 144–160. New York: Routledge.

———. 1992b. "From Secrets of Life to Secrets of Death." In *Secrets of Life, Secrets of Death: Essays on Language, Gender and Science*, 39–55. New York: Routledge.

———. 1992c. "Language and Ideology in Evolutionary Theory, Part I: Reading Cultural Norms into Natural Law." In *Secrets of Life, Secrets of Death: Essays on Language, Gender and Science*, 113–127. New York: Routledge.

———. 1992d. "Language and Ideology in Evolutionary Theory, Part II: The Language of Reproductive Autonomy." In *Secrets of Life, Secrets of Death: Essays on Language, Gender and Science*, 128–143. New York: Routledge.

———. 1995. *Refiguring Life: Changing Metaphors of Twentieth-Century Biology*. New York: Columbia University Press.

———. 1996. "Is There an Organism in this Text?" Paper presented at Growing Explanations: Historical Perspectives on Recent Scientific Practice, Princeton University, February 10.

Keller, Evelyn Fox, and Elisabeth A. Lloyd. 1992. "Introduction." In *Keywords in Evolutionary Biology*, ed. Evelyn Fox Keller and Elisabeth A. Lloyd, 1–6. Cambridge, Mass.: Harvard University Press.

Kelly, Kevin. 1991. "Designing Perpetual Novelty: Selected Notes from the Second Artificial Life Conference." In *Doing Science: The Reality Club*, ed. John Brockman, 1–44. New York: Prentice Hall.

———. 1994. *Out of Control: The Rise of Neo-Biological Civilization*. Redwood City, Calif.: Addison-Wesley.

Kevles, Daniel. 1977. *The Physicists: The History of a Scientific Community in Modern America*. New York: Vintage.

———. 1985. *In the Name of Eugenics: Genetics and the Uses of Human Heredity*. Berkeley: University of California Press.

Kitano, Hiroaki. 1997. "Towards Evolvable Electro-Biochemical Systems." In *Artificial Life V*, ed. Christopher G. Langton and Katsunori Shimohara, 219–226. Cambridge, Mass.: MIT Press/Bradford Books.

Knorr-Cetina, Karin D. 1981. *The Manufacture of Knowledge: An Essay on the Constructivist and Contextual Nature of Science*. Oxford: Pergamon Press.

———. 1983. "The Ethnographic Study of Scientific Work: Towards a Constructivist Interpretation of Science." In *Science Observed: Perspectives on the Social Study of Science*, ed. Karin Knorr-Cetina and Michael Mulkay, 115–140. London: Sage Publications.

Koza, John. 1992. *Genetic Programming: On the Programming of Computers by Means of Natural Selection*. Cambridge, Mass.: MIT Press.

———. 1994. "Artificial Life: Spontaneous Emergence of Self-Replicating and Evolutionary Self-Improving Computer Programs." In *Artificial Life II*, ed. Christopher G. Langton, Charles Taylor, J. Doyne Farmer, and Steen Rasmussen, 225–262. Redwood City, Calif.: Addison-Wesley.

Kristeva, Julia. 1983. "Stabat Mater." In *Tales of Love*, 234–263. Translated from the French by Leon S. Roudiez. New York: Columbia University Press, 1987.

Kuhn, Thomas. 1962. *The Structure of Scientific Revolutions*. Chicago: University of Chicago Press.

Laing, Richard. 1989. "Artificial Organisms: History, Problems, and Directions." In *Artificial Life*, ed. Christopher G. Langton, 49–62. Redwood City, Calif.: Addison-Wesley.

Lakoff, George, and Mark Johnson. 1980. *Metaphors We Live By*. Chicago: University of Chicago Press.

La Mettrie, Julien Offray de. 1748. *Man a Machine*. Translated from the French by M. W. Calkins. La Salle, Ill.: Open Court, 1988.

Lange, Marc. 1996. "Life, 'Artificial Life,' and Scientific Explanation." *Philosophy of Science* 63 (June): 225–244.

Langton, Christopher G. n.d. Letter to Dr. Stini. Undated correspondence, probably early 1980s. Courtesy Christopher G. Langton.

———. 1984. "Self-Reproduction in Cellular Automata." *Physica D* 10(1–2): 135–144.

———. 1988. "Toward Artificial Life." *Whole Earth Review* 58:74–79.

———. 1989. "Artificial Life." In *Artificial Life*, ed. Christopher G. Langton, 1–47. Redwood City, Calif.: Addison-Wesley.

———. 1991. "SimLife from Maxis: Playing with Virtual Nature." *Bulletin of the Santa Fe Institute* 7(2):4–6.

———. 1992. "Introduction." In *Artificial Life II*, ed. Christopher G. Langton, Charles Taylor, J. Doyne Farmer, and Steen Rasmussen, 3–24. Redwood City, Calif.: Addison-Wesley.

———. 1994. "From the Land of Enchantment to the Land of the Rising Sun: Artificial Life Spreads East." *Bulletin of the Santa Fe Institute* 9(1):9–12.

Latour, Bruno. 1983. "Give Me a Laboratory and I Will Raise the World." In *Science Observed: Perspectives on the Social Study of Science*, ed. Karin Knorr-Cetina and Michael Mulkay, 141–170. London: Sage Publications.

———. 1987. *Science in Action: How to Follow Scientists and Engineers through Society*. Cambridge, Mass.: Harvard University Press.

———. 1991. *We Have Never Been Modern*. Translated from the French by Catherine Porter. Cambridge, Mass.: Harvard University Press, 1993.

Latour, Bruno, and Steve Woolgar. 1986. *Laboratory Life: The Construction of Scientific Facts*, 2d ed. Princeton: Princeton University Press. first edition Beverly Hills: Sage Publications, 1979.

LeBaron, Blake. 1993. "The SFI Approach." In *Emergent Structures: A Newsletter of the Economics Research Program at the Santa Fe Institute*, 1–2. Santa Fe, New Mex.: Santa Fe Institute.

Le Guin, Ursula K. 1976. *The Word for World Is Forest*. New York: Berkeley.

———. 1989. "Do-It-Yourself Cosmology." In *The Language of the Night: Essays on Fantasy and Science Fiction*, 118–122. New York: HarperCollins.

Lem, Stanislaw. 1967. *The Cyberiad: Fables for the Cybernetic Age*. Translated from the Polish by Michael Kandel. San Diego: Harcourt Brace Jovanovich, 1985.

Lennox, James G. 1992. "Teleology." In *Keywords in Evolutionary Biology*, ed. Evelyn Fox Keller and Elisabeth A. Lloyd, 324–333. Cambridge, Mass.: Harvard University Press.

Lestel, D., L. Bec and J.-L. Lemoigne. 1993. "Visible Characteristics of Living Systems: Esthetics and Artificial Life." In *ECAL 93: Self-Organisation and*

Life: From Simple Rules to Global Complexity, ed. J. L. Deneubourg, S. Goss, G. Nicolis, H. Bersini, and R. Dagonnier, 595–603. Photocopied proceedings from the Second European Conference on Artificial Life.

Levy, Steven. 1991. "It's Alive!" *Rolling Stone,* June 13, 89, 91–92.

———. 1992a. "A-Life Nightmare." *Whole Earth Review* 76:34–47.

———. 1992b. *Artificial Life: The Quest for a New Creation.* New York: Pantheon.

Lewin, Roger. 1992. *Complexity: Life at the Edge of Chaos.* New York: Macmillan.

Lewontin, Richard C. 1984. "Adaptation." In *Conceptual Issues in Evolutionary Biology,* ed. Elliott Sober, 235–251. Cambridge, Mass.: MIT Press.

Lewontin, Richard C., Steven Rose, and Leon J. Kamin. 1984. *Not in Our Genes: Biology, Ideology, and Human Nature.* New York: Pantheon.

Lindenmayer, Aristid, and Przemyslaw Prusinkiewicz. 1989. "Developmental Models of Multicellular Organisms: A Computer Graphics Perspective." In *Artificial Life,* ed. Christopher G. Langton, 221–249. Redwood City, Calif.: Addison-Wesley.

Little, Linda. 1993. "Computational Platforms: Setting the Stage for Simulation." *Bulletin of the Santa Fe Institute* 8(2):13–16.

Lovelock, James. 1988. *The Ages of Gaia: A Biography of Our Living Earth.* New York: W. W. Norton.

Lutz, Catherine A., and Jane L. Collins. 1993. *Reading National Geographic.* Chicago: University of Chicago Press.

McIntosh, Robert. 1992. "Competition: Historical Perspectives." In *Keywords in Evolutionary Biology,* ed. Evelyn Fox Keller and Elisabeth A. Lloyd, 61–67. Cambridge, Mass.: Harvard University Press.

MacKenzie, Gordene Olga. 1994. *Transgender Nation: Transsexual Ideology, Transgenderism, and the Gender Movement in America.* Bowling Green: Bowling Green State University Popular Press.

MacLennan, Bruce. 1992. "Synthetic Ethology: An Approach to the Study of Communication." In *Artificial Life II,* ed. Christopher G. Langton, Charles Taylor, J. Doyne Farmer, and Steen Rasmussen, 631–658. Redwood City, Calif.: Addison-Wesley.

McMullin, Barry, and Francisco Varela. 1997. "Rediscovering Computational Autopoiesis." SFI preprint 97-02-012.

Macpherson, C. B. 1962. *The Political Theory of Possessive Individualism: Hobbes to Locke.* Oxford: Oxford University Press.

Malkki, Liisa. 1992. "National Geographic: The Rooting of Peoples and the Territorialization of National Identity among Scholars and Refugees." *Cultural Anthropology* 7:24–44.

Mandel, Ernest. 1972. *Late Capitalism.* Translated from the German by Joris De Bres. London: New Left Books, 1975.

Marcus, George E., and Michael M. J. Fischer. 1986. *Anthropology as Cultural Critique: An Experimental Moment in the Human Sciences.* Chicago: University of Chicago Press.

Marcus, George E., with Peter Dobkin Hall. 1992. *Lives in Trust: The Fortunes of Dynastic Families in Late Twentieth-Century America.* Boulder, Colo.: Westview Press.

Martin, Emily. 1991. "The Egg and the Sperm: How Science Has Constructed a Romance Based on Stereotypical Male-Female Roles." *Signs* (16)31:485–501.
———. 1994. *Flexible Bodies: Tracking Immunity in American Culture from the Days of Polio to the Age of AIDS.* Boston: Beacon.
Marx, Leo. 1964. *The Machine in the Garden: Technology and the Pastoral Ideal in America.* New York: Oxford University Press.
Maturana, Humberto, and Francisco Varela. 1980. *Autopoiesis and Cognition: The Realization of the Living.* Dordrecht: Reidel.
Maurer, Bill. 1992. "Can A Computer Understand? Hermeneutics in Computer Science." Manuscript.
———. 1995. "Complex Subjects: Offshore Finance, Complexity Theory, and the Dispersion of the Modern." *Socialist Review* 25(3–4):113–145.
Mayr, Ernst. 1976. *Evolution and the Diversity of Life: Selected Essays.* Cambridge, Mass.: Harvard University Press.
Merchant, Carolyn. 1980. *The Death of Nature: Women, Ecology, and the Scientific Revolution.* New York: Harper and Row.
Miller, Geoffrey F. and Peter M. Todd. 1993. "Evolutionary Wanderlust: Sexual Selection with Directional Mate Preferences." In *From Animals to Animals 2: Proceedings of the Second International Conference on Simulation of Adaptive Behavior,* ed. Jean-Arcady Meyer, Herbert L. Roitblat, and Stewart W. Wilson, 21–30. Cambridge, Mass.: MIT Press/Bradford Books.
Miller, Geoffrey, F., Peter M. Todd, and Shailesh U. Hegde. 1989. "Designing Neural Nets Using Genetic Algorithms." In *Proceedings of the Third International Conference on Genetic Algorithms,* ed. J. David Schaffer, 379–384. San Mateo, Calif.: Morgan Kaufmann.
Mitchell, Melanie. 1996. *An Introduction to Genetic Algorithms.* Cambridge, Mass.: MIT Press/Bradford Books.
Mitchell, Melanie, James Crutchfield, and Peter Hraber. 1993. "Dynamics, Computation, and the 'Edge of Chaos': A Re-Examination." SFI preprint 93-06-040.
Miura, Hirofumi, Takashi Yasuda, Yayoi Kubo Fujisawa, Yoshihiko Kuwana, Shoji Takeuchi, and Isao Shimoyama. 1997. "Insect-Model Based Microrobot." In *Artificial Life V,* ed. Christopher G. Langton and Katsunori Shimohara, 26–30. Cambridge, Mass.: MIT Press/Bradford Books.
Mohanty, Chandra Talpade. 1984. "Under Western Eyes: Feminist Scholarship and Colonial Discourse." *Boundary* 2–3(12–13):333–358.
Moravec, Hans. 1988. *Mind Children: The Future of Robot and Human Intelligence.* Cambridge, Mass.: Harvard University Press.
Morowitz, Harold. 1992. "Artificial Life: New Approaches to Theoretical Biology." *Bulletin of the Santa Fe Institute* 7(2):14–15.
Nader, Laura. 1974. "Up the Anthropologist: Perspectives Gained from Studying Up." In *Reinventing Anthropology,* ed. Dell Hymes, 284–311. New York: Vintage.
Nader, Laura, ed. 1996. *Naked Science: Anthropological Inquiry into Boundaries, Power, and Knowledge.* New York: Routledge.
Noble David. 1993. *A World without Women: The Christian Clerical Culture of Western Science.* Oxford: Oxford University Press.

———. 1997. *The Religion of Technology: The Divinity of Man and the Spirit of Invention.* New York: Knopf.

Nostrand, Richard L. 1992. *The Hispano Homeland.* Norman: University of Oklahoma Press.

Nye, David. 1994. *American Technological Sublime.* Cambridge, Mass.: MIT Press.

Obenzinger, Hilton. 1994. "American Palestine: Mark Twain, Herman Melville and Other American Travelers to the Holy Land, 1610–1882." Dissertation overview presented at the Stanford Humanities Center, Stanford University. October.

Ong, Aihwa. 1987. *Spirits of Resistance and Capitalist Discipline: Factory Workers in Malaysia.* Albany: State University of New York Press.

Otto, Rudolf. 1950. *The Idea of the Holy: An Inquiry into the Non-rational Factor in the Idea of the Divine and Its Relation to the Rational,* 2d ed. Translated from the German by John W. Harvey. London: Oxford University Press. First English edition 1923; first German edition 1917.

Oyama, Susan. 1985. *The Ontogeny of Information: Developmental Systems and Evolution.* Cambridge: Cambridge University Press.

Packard, Norman. 1989. "Intrinsic Adaptation in a Simple Model for Evolution." In *Artificial Life,* ed. Christopher G. Langton, 141–155. Redwood City, Calif.: Addison-Wesley.

Pateman, Carole. 1988. *The Sexual Contract.* Stanford: Stanford University Press.

Paxson, Heather. 1992. "Motherhood, Metaphor, and the Cultural Logic of Contract Pregnancy." Paper presented at Shifting Boundaries: The Eleventh Annual Lewis and Clark College Gender Studies Symposium, Lewis and Clark College, Portland, April 12–15.

Pfaffenberger, Bryan. 1988. "The Social Meaning of the Personal Computer: Or, Why the Personal Computer Revolution Was No Revolution." *Anthropological Quarterly* 61(1):39–47.

Pirsig, Robert M. 1974. *Zen and the Art of Motorcycle Maintenance: An Inquiry into Values.* New York: William Morrow.

Prediction Company. 1993. *Corporate Overview.* Santa Fe, New Mex.: Prediction Company.

Putnam, Jeffrey. 1993. "A Primordial Soup Environment." In *ECAL 93: Self-Organisation and Life: From Simple Rules to Global Complexity,* ed. J. L. Deneubourg, S. Goss, G. Nicolis, H. Bersini, and R. Dagonnier, 943–961. Photocopied proceedings from the Second European Conference on Artificial Life.

Rabinow, Paul. 1992. "Artificiality and Enlightenment: From Sociobiology to Biosociality." In *Incorporations,* ed. Jonathan Crary and Sanford Kwinter, 234–252. New York: Zone.

———. 1996. *Making PCR: A Story of Biotechnology.* Chicago: University of Chicago Press.

Rappaport, Roy. 1967. *Pigs for the Ancestors.* New Haven: Yale University Press.

Rasmussen, Steen. 1992. "Aspects of Information, Life, Reality, and Physics." In *Artificial Life II,* ed. Christopher G. Langton, Charles Taylor, J. Doyne Farmer, and Steen Rasmussen, 767–773. Redwood City, Calif.: Addison-Wesley.

Rasmussen, Steen, Rasmus Feldberg, and Carsten Knudsen. 1992. "Self-Programming of Matter and the Evolution of Proto-Biological Organizations." SFI preprint 92-07-035.

Rawlins, Gregory J. E. 1997. *Slaves of the Machine: The Quickening of Computer Technology.* Cambridge, Mass.: MIT Press/Bradford Books.

Ray, Thomas S. 1991. "Is It Alive or Is It GA?" In *Proceedings of the Fourth International Conference on Genetic Algorithms,* ed. Richard K. Belew and Lashon B. Booker, 527–535. San Mateo, Calif.: Morgan Kaufmann.

———. 1992a. "An Approach to the Synthesis of Life." In *Artificial Life II,* ed. Christopher G. Langton, Charles Taylor, J. Doyne Farmer, and Steen Rasmussen, 371–408. Redwood City, Calif.: Addison-Wesley.

———. 1992b. "Natural Evolution of Machine Codes: Digital Organisms." In *SFI Proposal for a Research Program in Adaptive Computation.* Santa Fe, New Mex.: Santa Fe Institute.

———. 1994a. "An Evolutionary Approach to Synthetic Biology: Zen and the Art of Creating Life." *Artificial Life* 1(1–2):179–210.

———. 1994b. "A Proposal to Create a Network-Wide Biodiversity Reserve for Digital Organisms." Paper presented at Artificial Life IV, MIT, Mass., July 6–8.

Regis, Ed. 1990. *Great Mambo Chicken and the Transhuman Condition: Science Slightly over the Edge.* Redwood City, Calif.: Addison-Wesley.

Reynolds, Craig. 1987. "Flocks, Herds, and Schools: A Distributed Behavioral Model." (Proceedings of SIGGRAPH 1987) *Computer Graphics* 21(4):25–34.

———. 1992. "Boids Demo." *Artificial Life II Video Proceedings,* ed. Christopher G. Langton. Redwood City, Calif.: Addison-Wesley.

———. 1994. "Competition, Coevolution, and the Game of Tag." In *Artificial Life IV,* ed. Rodney Brooks and Pattie Maes, 59–69. Cambridge, Mass.: MIT Press.

Reynolds, Peter C. 1993. "The Priests of Cyborg." *Month,* July, 257–266.

Risan, Lars. 1996. "Artificial Life: A Technoscience Leaving Modernity? An Anthropology of Subjects and Objects." Cand. polit. dissertation, University of Oslo.

Roediger, David. 1994. *Towards the Abolition of Whiteness: Essays on Race, Politics, and Working Class History.* London: Verso.

Rogers, Ada. 1995. "Is There a Case for Viruses?" *Newsweek,* February 27, 65.

Ross, Andrew. 1991. "New Age Technoculture." In *Cultural Studies,* ed. Lawrence Grossberg, Gary Nelson, and Paula Triechler, 531–555. New York: Routledge.

———. 1995. "The Great White Dude." In *Constructing Masculinity,* ed. Maurice Berger, Brian Wallis, and Simon Watson, 167–175. New York: Routledge.

Rothschild, Michael. 1990. *Bionomics: The Inevitability of Capitalism.* New York: Holt.

Rotman, Brian. 1993. *Ad Infinitum—The Ghost in Turing's Machine: Taking God out of Mathematics and Putting the Body back in: An Essay in Corporeal Semiotics.* Stanford: Stanford University Press.

Rucker, Rudy, R. U. Sirius, and Queen Mu, eds. 1992. *Mondo 2000: A User's Guide to the New Edge.* New York: HarperCollins.

Russ, Joanna. 1975. *The Female Man.* New York: Bantam.

Sagan, Dorion. 1992. "Metametazoa: Biology and Multiplicity." In *Incorporations,* ed. Jonathan Crary and Sanford Kwinter, 362–385. New York: Zone.

Santa Fe Institute. 1991. *Bulletin of the Santa Fe Institute* 6(2).

———. 1993. *Simple Rules . . . Complex Behavior.* Promotional video for the Santa Fe Institute.

———. 1994a. *Brochure for the New Campus of the Santa Fe Institute.* Santa Fe, New Mex.: Santa Fe Institute.

———. 1994b. *1993 Annual Report on Scientific Programs.* Santa Fe, New Mex.: Santa Fe Institute.

———. 1996. *Bulletin of the Santa Fe Institute* 11(1).

———. 1997. *Bulletin of the Santa Fe Institute* 12(1).

Sardar, Ziauddin. 1996. "alt.civilizations.faq: Cyberspace as the Darker Side of the West." In *Cyberfutures: Culture and Politics on the Information Superhighway,* ed. Ziauddin Sardar and Jerome R. Ravetz, 14–41. New York: New York University Press.

Schaffer, Simon. 1994. "Babbage's Intelligence: Calculating Engines and the Factory System." *Critical Inquiry* 21(1):203–227.

Schneider, David. 1968. *American Kinship: A Cultural Account.* Chicago: University of Chicago Press.

Schrödinger, Erwin. 1944. *What Is Life? The Physical Aspect of the Living Cell.* Cambridge: Cambridge University Press.

Schwartz, Hillel. 1996. *The Culture of the Copy: Striking Likenesses, Unreasonable Facsimilies.* New York: Zone.

Searle, John R. 1980. "Minds, Brains, Programs." *Behavioral and Brain Sciences* 3:417–424.

Shelley, Mary. 1818. *Frankenstein, or the Modern Prometheus.* Harmondsworth: Penguin, 1985.

Sims, Karl. 1992. "Panspermia." *Artificial Life II Video Proceedings,* ed. Christopher G. Langton. Redwood City, Calif.: Addison-Wesley.

———. 1994. "Evolving 3D Morphology and Behavior by Competition." *Artificial Life* 1(4):353–372.

Skipper, Jakob. 1992. "The Computer Zoo—Evolution in a Box." In *Toward a Practice of Autonomous Systems: Proceedings of the First European Conference on Artificial Life,* ed. Francisco J. Varela and Paul Bourgine, 355–364. Cambridge, Mass.: MIT Press/Bradford Books.

Smith, Brian Cantwell. 1994. "Coming Apart at the Seams: The Role of Computation in a 'Successor' Metaphysics." Paper presented at Vital Signs: Cultural Perspectives on Coding Life and Vitalizing Code, Stanford University, June 2–4.

Smith, Neil. 1996. "The Production of Nature." In *FutureNatural: Nature, Science, Culture,* ed. George Robertson, Melinda Mash, Lisa Tucker, Jon Bird, Barry Curtis, and Tim Putnam, 35–54. London: Routledge.

Smith, Robert E., Stephanie Forrest, and Alan S. Perelson. 1992. "Searching for Diverse, Cooperative Populations with Genetic Algorithms." Santa Fe Institute preprint 92-06-027.

Smyth, Cherry. 1995. "How Shall I Address You? Pronouns, Pussies, and Pricks—Talking to Female-to-Male Transsexuals." *On Our Backs* 11(1): 18–41.

Sobchack, Vivian. 1993. "New Age Mutant Ninja Hackers: Reading *Mondo 2000*," *South Atlantic Quarterly* 92(4):569–584.

Sober, Elliott. 1992. "Learning from Functionalism—Prospects for Strong Artificial Life." In *Artificial Life II*, ed. Christopher G. Langton, Charles Taylor, J. Doyne Farmer, and Steen Rasmussen, 749–765. Redwood City, Calif.: Addison-Wesley.

Spafford, Eugene. 1994. "Computer Viruses as Artificial Life." *Artificial Life* 1(3):249–266.

Spivak, Gayatri Chakravorty. 1985. "Scattered Speculations on the Question of Value." In *In Other Worlds: Essays in Cultural Politics*, 154–175. London: Routledge, 1988.

Stengers, Isabelle. 1995. "God's Heart and the Stuff of Life." Paper presented at the Third European Conference on Artificial Life, Granada, Spain, June 4–6.

Stepan, Nancy Leys. 1993. "Race and Gender: The Role of Analogy in Science." In *The "Racial" Economy of Science: Toward a Democratic Future*, ed. Sandra Harding, 359–376. Bloomington: Indiana University Press.

Stone, Allucquère Rosanne. 1995. *The War of Desire and Technology at the Close of the Mechanical Age.* Cambridge, Mass.: MIT Press.

Strathern, Marilyn. 1992a. *After Nature: English Kinship in the Late Twentieth Century.* Cambridge: Cambridge University Press.

———. 1992b. *Reproducing the Future: Anthropology, Kinship, and the New Reproductive Technologies.* New York: Routledge.

Sturgeon, Theodore. 1941. "Microcosmic God." *Astounding Science-Fiction,* April.

Taussig, Michael. 1993. *Mimesis and Alterity: A Particular History of the Senses.* New York: Routledge,

Taylor, Charles, David Jefferson, Scott Turner, and Seth Goldman. 1989. "RAM: Artificial Life for the Exploration of Complex Biological Systems." In *Artificial Life,* ed. Christopher G. Langton, 275–296. Redwood City, Calif.: Addison-Wesley.

Taylor, Peter, Saul Halfon, and Paul Edwards, eds. 1997. *Changing Life: Genomes, Ecologies, Bodies, Commodities.* Minneapolis: University of Minnesota Press.

Terzopoulos, Demetri, Tamer Rabie, and Radek Grzeszczuk. 1997. "Perception and Learning in Artificial Animals." In *Artificial Life V,* ed. Christopher G. Langton and Katsunori Shimohara, 346–353. Cambridge, Mass.: MIT Press/Bradford Books.

Terzopoulos, Demetri, Xiaoyuan Tu, and Radek Grzeszczuk. 1994. "Artificial Fishes: Autonomous Locomotion, Perception, Behavior, and Learning in a Simulated Physical World." *Artificial Life* 1(4):327–351.

Theweleit, Klaus. 1977. *Male Fantasies. Vol. 1: Women, Floods, Bodies, History.* Translated from the German by Stephen Conway in collaboration with Erica Carter and Chris Turner. Minneapolis: University of Minnesota Press, 1987.

Thompson, D'Arcy. 1917. *On Growth and Form.* Cambridge: Cambridge University Press.

Todd, Peter. 1993. "Artificial Death." In *ECAL 93: Self-Organisation and Life: From Simple Rules to Global Complexity,* ed. J. L. Deneubourg, S. Goss, G. Nicolis, H. Bersini, and R. Dagonnier, 1048–1059. Photocopied proceedings from the Second European Conference on Artificial Life.

Todd, Peter M., and Geoffrey F. Miller. 1991. "On the Sympatric Origin of Species: Mercurial Mating in the Quicksilver Model." In *Proceedings of the Fourth International Conference on Genetic Algorithms,* ed. Richard K. Belew and Lashon B. Booker, 547–554. San Mateo, Calif.: Morgan Kaufmann.

———. 1997. "Biodiversity through Sexual Selection." In *Artificial Life V,* ed. Christopher G. Langton and Katsunori Shimohara, 289–299. Cambridge, Mass.: MIT Press/Bradford Books.

Tosa, Naoko, and Ryohei Nakatsu. 1997. "The Esthetics of Artificial Life." In *Artificial Life V,* ed. Christopher G. Langton and Katsunori Shimohara, 143–151. Cambridge, Mass.: MIT Press/Bradford Books.

Travers, Michael. 1989. "Animal Construction Kits." In *Artificial Life,* ed. Christopher G. Langton, 421–442. Redwood City, Calif.: Addison-Wesley.

Traweek, Sharon. 1988. *Beamtimes and Lifetimes: The World of High Energy Physicists.* Cambridge, Mass.: Harvard University Press.

Trinh T. Minh-ha, ed. 1986–1987. "Difference: 'A Special Third World Women's Issue.' " *Discourse: Journal for Theoretical Studies in Media and Culture* 8:3–38.

Tsing, Anna Lowenhaupt. 1995. "Empowering Nature, or: Some Gleanings in Bee Culture." In *Naturalizing Power: Essays in Feminist Cultural Analysis,* ed. Sylvia Yanagisako and Carol Delaney, 113–143. New York: Routledge.

Tuana, Nancy. 1989. "The Weaker Seed: The Sexist Bias of Reproductive Theory." In *Feminism and Science,* ed. Nancy Tuana, 147–171. Bloomington: Indiana University Press.

Turing, Alan. 1950. "Computing Machinery and Intelligence." *Mind* 59(236): 433–460. Reprinted as "Can a Machine Think?" In *The World of Mathematics,* vol. 4, ed. James R. Newman, 2099–2123. New York: Simon and Schuster, 1956.

Turkle, Sherry. 1991. "Romantic Reactions: Paradoxical Responses to the Computer Presence." In *Boundaries of Humanity: Humans, Animals, Machines,* ed. James J. Sheehan and Morton Sosna, 224–252. Berkeley: University of California Press.

———. 1995. *Life on the Screen: Identity in the Age of the Internet.* New York: Simon and Schuster.

Varela, Francisco. 1979. *Principles of Biological Autonomy.* New York: Elsevier North-Holland.

———. 1992. "The Reenchantment of the Concrete." In *Incorporations,* ed. Jonathan Crary and Sanford Kwinter, 320–338. New York: Zone.

————. 1995. "Heinz von Foerster, the Scientist, the Man: Prologue to the Interview." *Stanford Humanities Review* 4(2):285–287.

Varela, Francisco, Evan Thompson, and Eleanor Rosch. 1991. *The Embodied Mind: Cognitive Science and Human Experience*. Cambridge, Mass.: MIT Press.

Varela, Francisco, Humberto Maturana, and R. Uribe. 1974. "Autopoiesis: The Organization of Living Systems, Its Characterization and a Model." *BioSystems* 5:187–196.

Vonnegut Kurt. 1963. *Cat's Cradle*. New York: Dell, 1970.

von Neumann, John. 1966. *Theory of Self-Reproducing Automata*. Edited by Arthur Burks. Urbana: University of Illinois Press.

Waldrop, M. Mitchell. 1992. *Complexity: The Emerging Science at the Edge of Order and Chaos*. New York: Simon and Schuster.

Walsh, Birrell. 1994. "The Coming Convergence of Biology and Technology: An Interview with Kevin Kelly." *Microtimes*, November 14, 214–330.

Warner, Michael. 1993. "Introduction." In *Fear of a Queer Planet: Queer Politics and Social Theory*, ed. Michael Warner, vi–xxxi. Minneapolis: University of Minnesota Press.

Werner, Gregory. 1997. "Why the Peacock's Tail Is So Short: Limits to Sexual Selection." In *Artificial Life V*, ed. Christopher G. Langton and Katsunori Shimohara, 85–91. Cambridge, Mass.: MIT Press/Bradford Books.

Werner, Gregory, and Michael Dyer. 1992. "Evolution of Communication in Artificial Organisms." In *Artificial Life II*, ed. Christopher G. Langton, Charles Taylor, J. Doyne Farmer, and Steen Rasmussen, 659–688. Redwood City, Calif.: Addison-Wesley.

Weston, Kathleen. 1991. *Families We Choose: Lesbians, Gays, Kinship*. New York: Columbia University Press.

Wheeler, Michael. 1995. "Escaping from the Cartesian Mind-Set: Heidegger and Artificial Life." In *Advances in Artificial Life: The Third European Conference on Artificial Life* (Lecture Notes in Artificial Intelligence 929, Subseries of Lecture Notes in Computer Science), ed. F. Morán, A. Moreno, J. J. Morelo, and P. Chacón, 65–76. Berlin: Springer.

————. 1996. "From Robots to Rothko: The Bringing Forth of Worlds." In *The Philosophy of Artificial Life*, ed. Margaret A. Boden, 209–236. New York: Oxford University Press.

Wiener, Norbert. 1948. *Cybernetics, or Control and Communication in the Animal and the Machine*. New York: Wiley.

Williams, Raymond. 1977. "Hegemony." In *Marxism and Literature*, 108–114. Oxford: Oxford University Press.

Wilson, Chris. 1997. *The Myth of Santa Fe: Constructing a Modern Regional Tradition*. Albuquerque: University of New Mexico Press.

Wilson, Edward O. 1975. *Sociobiology: The New Synthesis*. Cambridge, Mass.: Belknap/Harvard University Press.

Winograd, Terry. 1972. *Understanding Natural Language*. New York: Academic Press.

Winterson, Jeanette. 1992. *Written on the Body*. London: Jonathan Cape.

Wittig, Monique. 1992. "The Category of Sex." In *The Straight Mind and Other Essays,* 1–8. Boston: Beacon.

Wolf, Eric R. 1982. *Europe and the People without History.* Berkeley: University of California Press.

Wright, Robert. 1988. *Three Scientists and Their Gods: Looking for Meaning in an Age of Information.* New York: Harper and Row.

Wuensche, Andrew. 1994. "Complexity in One-D Cellular Automata: Gliders, Basins of Attraction and the Z Parameter." SFI preprint 94-04-025.

Yaeger, Larry. 1994. "Computational Genetics, Physiology, Metabolism, Neural Systems, Learning, Vision, and Behavior or PolyWorld: Life in a New Context." In *Artificial Life III,* ed. Christopher G. Langton, 263–298. Redwood City, Calif.: Addison-Wesley.

Yanagisako, Sylvia. 1978. "Variance in American Kinship: Implications for Cultural Analysis." *American Ethnologist* 5(1):15–29.

Yanagisako, Sylvia, and Carol Delaney. 1995. "Naturalizing Power." In *Naturalizing Power: Essays in Feminist Cultural Analysis,* ed. Sylvia Yanagisako and Carol Delaney, 1–22. New York: Routledge.

Index

Compositor:	Bookmasters, Inc.
Text:	Sabon
Display:	Sabon
Printer:	Data Reproductions
Binder:	John H. Dekker and Sons